EUCLID IN GREEK

EUCLID IN GREEK

BOOK I

WITH INTRODUCTION
AND NOTES

BY

SIR THOMAS L. HEATH

K.C.B., K.C.V.O., F.R.S.

SC.D. CAMB.

HON. D.SC. OXFORD

CAMBRIDGE
AT THE UNIVERSITY PRESS
1920

CAMBRIDGE UNIVERSITY PRESS
Cambridge, New York, Melbourne, Madrid, Cape Town, Singapore,
São Paulo, Delhi, Dubai, Tokyo, Mexico City

Cambridge University Press
The Edinburgh Building, Cambridge CB2 8RU, UK

Published in the United States of America by
Cambridge University Press, New York

www.cambridge.org
Information on this title: www.cambridge.org/9780521183475

First published 1920
First paperback edition 2010

A catalogue record for this publication is available from the British Library

ISBN 978-0-521-18347-5 Paperback

PREFACE

IN these days when Greek is supposed to be on its trial and Euclid happily defunct, it may well seem a wildly reactionary proceeding to suggest to teachers a combination of the two, a piling (so it might be thought) of one inutility on another. But, first, we must bear in mind that it is only compulsory Greek that is threatened: when that is gone, the study of Greek will be no whit less necessary to a complete education. Generation after generation of men and women will still have to go to school to the Greeks for the things in which they are our masters; and for this purpose they must continue to learn Greek. Again, Euclid can never at any time be more than apparently in abeyance; he is immortal. Elementary geometry will also continue to form part of a complete education; and elementary geometry *is* Euclid, however much the editors of text-books may try to obscure the fact.

But I am not here concerned to argue the case of Euclid against other text-books of geometry. The aim of this book is to maintain an opinion which I have long held that, if the study of Greek and Euclid be combined by reading at least part of Euclid in the original, the two elements will help each other enormously. In the first place, boys

learning Greek in the higher Forms in schools will
generally have some knowledge of elementary geometry. Even if this is read in some text-book other
than Euclid, all the technical terms and phrases will
be the same, and in any proposition of Euclid that
may be taken up the course of the proof will be
easily divined, almost by simple inspection. Hence,
in translating the Greek, the student will really be
translating something quite familiar. Now every
one knows that, when beginning the study of a
foreign language, it is quite a good plan to read
familiar chapters of the Bible in a translation into
the particular tongue. The advantage to the student
of Greek of reading in Greek the familiar propositions of Euclid's first book will hardly be less.
Secondly, from the point of view of learning geometry, much advantage will be gained by having,
as it were, to spell out the Greek. The beginner in
geometry needs to learn a good many things by
heart, especially technical terms, many of which,
being reflections of the Greek, will be the better
understood if the Greek forms are known. How can
any person who has only had such words as *theorem*,
problem, *isosceles*, *parallelepiped* explained to him in
English apart from their derivation get any such
clear idea of their significance as the person who
knows them as θεώρημα, πρόβλημα, ἰσοσκελές, παραλ-
ληλεπίπεδον? Again, persons with no particular
aptitude for mathematics find a difficulty in memor-
ising the course of the proofs and avoiding confusion

between propositions which are nearly allied, e.g. as converses. I cannot but think that the formality and deliberation of Euclid's exposition, involving a certain amount of repetition of stereotyped phrases or steps, combined with the necessity of translating the Greek word for word wherever they occur, would be a very effective way of impressing the technique and the whole content on the mind of the learner.

A last consideration is this. I remember well (it was in the days long ago when most people were under the impression that Books VII—X of Euclid were lost in Greek) the thrill of pleasure I felt when I first took up the Greek text of Euclid, then not too easy to get hold of, for there was no Teubner text as there is now, and one had to look in College libraries for copies, which in any case were not too convenient to handle. I cannot but think that at all events the studious boys in the higher Forms of schools who have already got a fair grasp of Greek and know the amount of geometry corresponding to the first book of Euclid would be really interested to see the actual language in which the old Alexandrian taught the youth and pupils of maturer age in his own day, and so to put themselves in the place of their fellow-students of twenty-two centuries ago. It is in the confident belief that there will be many senior boys at school and students at the Universities, to say nothing of students of riper years, who will in this way come to read Euclid with more zest and more profit that I offer them this little book.

It is desirable to add a word with regard to the notes. I am convinced that there is no subject which, if properly presented, is better calculated than the fundamentals of geometry to make the schoolboy (or the grown man) *think*. This is the object of the notes; and, if it is attained, it is worth while, whatever the schoolboy's after career may be.

T. L. H.

5. xi. 19

CONTENTS

INTRODUCTION

INTRODUCTION

EUCLID

ANYONE writing on Euclid would wish to be in a position to give some particulars of the life and personality of the man who wrote a great classic the reign of which, at least among the civilised nations of the West, can hardly have been paralleled by that of any book except the Bible.

Unfortunately however we have only the most meagre information about him. It was apparently the fate of men of science in Greece as elsewhere to make little noise in the world as such. A still greater man, Archimedes, only appears in history because he figured as a defender of Syracuse during the siege of that place by Marcellus in the Second Punic War; his mechanical appliances (engines of war) were a constant terror to the Romans, and it was these which made him famous, while he himself thought meanly of mechanics and all practical arts and devoted himself, heart and soul, to the theoretical investigation of abstruse problems in pure mathematics.

Euclid is claimed by Proclus as a Platonist, but all that can be assumed is that he received his mathematical training at Athens from the pupils of Plato; for most of the geometers who could have taught him belonged to that school, and it was in Athens that the older writers of Elements had lived and taught.

It is clear that he was in date intermediate between the first pupils of Plato (and, in particular, Eudoxus, who lived approximately from 408 to 355 B.C.) on the one hand and Archimedes on the other (for Archimedes quotes him textually). He is assigned to the reign of the first Ptolemy (306–283 B.C.), and he flourished probably about 300 B.C. or a little earlier. He taught and founded a school at Alexandria, for we are told that Apollonius of Perga, the author of the great treatise on *Conics*, spent a long time with the pupils of Euclid at Alexandria.

There are some stories of him which one would like to believe true. Stobaeus tells us that some one who had begun to read geometry with Euclid, when he had learnt the first proposition, inquired, 'What shall I get by learning these things?' whereon Euclid called his slave and said, 'Give him threepence, since he must needs make gain out of what he learns.'

In another story he appears as the instructor of Ptolemy in geometry. Ptolemy evidently, like some moderns, found the subject long-drawn-out, for he asked whether there was not any shorter way to geometry than that of the *Elements*, to which Euclid replied that there was no royal road to geometry.

The Arabs, who assimilated Greek geometry with avidity, have more circumstantial accounts of Euclid, to the effect, e.g., that he was the son of Naucrates and grandson of Zenarchus, a Greek born at Tyre and domiciled at Damascus. But the Arabian stories seem to be partly due to misunderstandings and partly invented in order to gratify a desire which the Arabs always showed of connecting famous Greeks in some way or other with the East (the same predilection made them describe Pythagoras as a pupil of the wise Salomo, Hipparchus as the exponent of Chaldaean philosophy or as the Chaldaean, Archimedes as an Egyptian, and so on). They even invented an explanation of the name of Euclid, which they variously pronounced as Uclides or Icludes, making it a compound of *Ucli*, a key, and *Dis*, a measure, or, as some say, geometry, so that Uclides is equivalent to the key of geometry!

WORKS OTHER THAN THE ELEMENTS

THE *Elements*, though the most famous, was only one of many treatises by Euclid. He seems indeed to have covered the whole field of mathematics as then known.

In enumerating the other known works of Euclid, we will begin with those which, like the *Elements*, were concerned with elementary geometry.

The *Pseudaria*, or *Fallacies*, which seems to be irretrievably lost, was a sort of foil to the *Elements*. In the efforts of geometers to discover new theorems or to solve new problems, many cases had arisen where the authors, although basing themselves on

the true principles of the science, had through
errors in reasoning or method deduced conclusions
which were false. Euclid's book seems to have been
a sort of guide to the causes of error, and to have
contained illustrations of the false deductions, with
practical hints to enable the beginner to detect the
flaw in the argument, to refute it, and, by dint of
practice in distinguishing the true from the false, to
avoid such errors in his own work.

The *Data* (δεδομένα) is fortunately preserved in
Greek, and the full text with the commentary by
Marinus has been newly edited by Menge (Heiberg
and Menge's *Euclidis opera omnia*, vol. vi, Teubner,
1896). A translation of the *Data* was also included in the later editions of Simson's Euclid
(though his text left much to be desired). The
form of the *Data* is that of propositions proving
that, if certain things in a figure are ' given ' (in
magnitude, in position, or in *species*, i.e. in shape)
something else is ' given,' that is to say, can be
actually determined. This collection of *Data*, convenient for reference, enabled the solution of small
subsidiary problems occurring in a larger investigation to be taken for granted instead of being
worked out on each occasion. We may, by way of
example, quote one enunciation, namely that of
the proposition which gives the equivalent of the
solution of a general quadratic equation. ' If two
straight lines contain a parallelogram given in
magnitude in a given angle, and if the sum of the

straight lines be given, then shall each of them be given'; that is, if $xy = b^2$, a given area, and $x + y = a$, a given length, then x and y are both given, i.e. can be determined.

Περὶ διαιρέσεων βιβλίον, a tract *On Divisions* (of figures), is lost in Greek but has been discovered in the Arabic. Woepcke found it in a MS. at Paris and published it with a translation in 1851. It is expressly attributed to Euclid and corresponds to the description of it by Proclus. There is no doubt that the tract edited by Woepcke is not only Euclid's own work but the whole of it. It has been restored and published with full commentary and introductory chapters by R. C. Archibald (Cambridge, 1915). The divisions of figures are divisions into other figures either like or unlike; thus a triangle is divided into triangles or into a triangle and a quadrilateral; a figure bounded by an arc of a circle and two straight lines drawn from a point to the extremities of the arc is divided into two equal parts; from a circle a given fraction of it is cut off between two parallel chords; and so on.

To higher geometry belong the following.

Three books of *Porisms*. These are lost, and all we know of them is contained in the *Collection* of Pappus. There have been several attempts at restoration, but none that can be regarded as disposing of the subject. This much is clear, that the *Porisms* belonged to higher geometry and contained propositions forming part of the modern theory of transversals and of projective geometry. It contained the basis of the theory of anharmonic ratios.

The *Surface-Loci* (τόποι πρὸς ἐπιφανείᾳ). This too is lost, and nothing is known of it except from Pappus, who mentions the book as part of the 'Treasury of Analysis' (τόπος ἀναλυόμενος) and gives two lemmas on it. It seems to have dealt with such loci as are cones, cylinders and spheres. One of Pappus's lemmas states and proves completely the focus-directrix property of the three conic sections, which may therefore be taken to have been assumed in Euclid's work as known.

The *Conics*. This treatise too is lost, having evidently been superseded by the great work of Apollonius about a century later. Pappus says of it, 'The four books of Euclid's Conics were completed by Apollonius, who added four more and gave us eight books of Conics.' It is probable that Euclid's work was already lost in Pappus's time, for he goes on to speak of 'Aristaeus who wrote the *still extant* five books of Solid Loci connected with the Conics.' 'Solid Loci' (στερεοὶ τόποι) in Greek terminology were actually conic sections. Probably Aristaeus's work was on conics regarded as loci, and Euclid's treatise, though general in scope like that of Apollonius, was confined to those properties which were necessary for the analysis of the *Solid Loci* of Aristaeus.

The *Phaenomena*. This is an astronomical work and is still extant. A much interpolated version appeared in Gregory's edition of Euclid, but it has now been edited by Menge from an earlier and better source (*Euclidis opera omnia*, vol. VIII, Teubner, 1916). The book consists of 16 (or 18) propositions of *sphaeric* geometry and is based partly on Autolycus's work *On the moving sphere*, and partly on an earlier text-book of *Sphaerica* of exclusively mathematical content.

The *Optics*. This book too survives and is included in Heiberg and Menge's edition (vol. VII, Teubner, 1895). The *Catoptrica* (theory of mirrors) included by Heiberg in the same volume is not by Euclid, and Heiberg suspects that in its present form it may be due to Theon of Alexandria (fourth century A.D.), the editor of the *Elements*.

Euclid is also said to have written on the Elements of Music. Two treatises are attributed to him in our MSS. of the *Musici*. One is the *Sectio Canonis* (κατατομὴ κανόνος), the theory of the musical intervals. Its genuineness is however doubtful. Jan, the editor of the *Musici*, thought it genuine, on two main grounds, (1) that the form and style agree well with what we find in the *Elements* and (2) that in an ancient commentary on Ptolemy's *Harmonica* Euclid is thrice mentioned as the author of the *Sectio*, and almost the whole of the treatise except the preface is quoted *in extenso*. On the other hand Tannery disputed the authenticity, and the latest editor, Menge, who finds in the tract a number of expressions and at all events one proof which are not in Euclid's manner, concludes that in its present form it is not Euclid's but is more likely to have been excerpted by some not too well equipped compiler from the fuller *Elements of Music* mentioned by Proclus and Marinus. The other treatise, an *Introduction to Harmonics* (εἰσαγωγὴ ἁρμονική), is not by Euclid but by Cleonides, a pupil of Aristoxenus. Both treatises are included in Heiberg and Menge's edition, vol. VIII (Teubner, 1916).

EARLIER WRITERS OF ELEMENTS

THE first writer of Elements was, we are told, Hippocrates of Chios (fl. in the second half of the fifth century B.C.), a great geometer, who was famous for two other achievements, (1) the reduction of the problem of doubling the cube to that of finding two mean proportionals in continued proportion between two given straight lines, (2) the quadrature of certain *lunes* which can be squared by means of the geometry of the straight line and circle. From the account of these quadratures we can judge to some extent how far the Elements had been developed up to Hippocrates's time; he is quite familiar with the main propositions of Euclid's Book III, and he is himself said to have proved that the areas of circles are to one another as the squares on their diameters (the theorem of Eucl. XII, 2). Next Leon, who came between Plato (429–348 B.C.) and Eudoxus (408–355 B.C.), put together a more careful collection, the propositions in it being at once more

numerous and more serviceable. The geometrical
text-book of the Academy was written by Theudius
of Magnesia, who, with Amyclas of Heraclea,
Menaechmus the pupil of Eudoxus, Dinostratus
the brother of Menaechmus, and Athenaeus of
Cyzicus, consorted together in the Academy and
carried on their investigations in common. Her-
motimus of Colophon is also said to have ' discovered
many of the elements,' but no text-book after that
of Theudius is mentioned. Theudius therefore must
be taken to be the immediate precursor of Euclid.
Euclid himself no doubt made full use of Theudius
as well as of all other available material. Euclid is
naturally silent as to his relation to his predecessor;
but we find in Aristotle (who was fond of geometrical illustrations) indications of proofs of certain
propositions, proofs which he must have taken from
some recognised text-book (probably that of Theudius) and which differ from Euclid's proofs. Some
light is thus thrown on the changes which Euclid
made in the methods of his predecessors.

CONTENTS OF EUCLID'S ELEMENTS

A SHORT indication of the contents of the *Elements*
will not be out of place. Book I is in three sections.
The first section deals mainly with triangles, their
construction and their properties in the sense of the
relation of their parts, the sides and angles, to one
another, and the comparison of different triangles

in respect of their parts, and of their area where they are congruent; it also treats of 'adjacent' and 'vertically opposite' angles made by two straight lines; and it contains a few simple problems of construction, the drawing of perpendiculars, and the bisection of a straight line and an angle respectively. The second section beginning with 1, 27 establishes the theory of parallels, leading to the proposition (1, 32) that the sum of the angles of any triangle is equal to two right angles. The third section begins with 1, 33, 34 which introduce the parallelogram for the first time, and this section of Book I with Book II carries the subject up to the transformation of rectilineal areas of any shape into parallelograms of any shape, rectangles, and ultimately squares. Book III consists of the elementary geometry of the circle, and Book IV solves the problems of inscribing in circles and circumscribing about circles (1) triangles of any shape, (2) the regular polygons which can be inscribed or circumscribed by means of the elementary geometry of the straight line and circle, namely the square, the pentagon, the hexagon, the decagon and the regular polygon of fifteen sides. The content of practically the whole of Books I, II, IV was Pythagorean; and although we know little of the Pythagorean contributions to the geometry of the circle, Book III also may well be to a great extent Pythagorean, as it is certainly pre-Euclidean.

Book V expounds the theory of proportion in its most general form, applicable to magnitudes of all kinds, both commensurable and incommensurable. The whole theory is due to Eudoxus.

Book VI applies the new theory of proportion expounded in Book V to plane geometry. Much of this Book is also Pythagorean and, in particular, the propositions dealing with the most general cases of the *application of areas* (Props. 27–29, which are equivalent to the solution of the most general forms of the quadratic equation in algebra so far as the roots are real), and the construction of a rectilineal figure equal to one given rectilineal figure and similar to another. The treatment of these questions is more perfect than the Pythagoreans could give, in so far as Euclid is able to use the perfected theory of proportion, whereas the Pythagorean theory of proportion was only applicable to commensurable magnitudes and consequently their geometrical investigations which involved the use of proportions were not, so to speak, watertight.

Books VII–IX deal with arithmetic in the sense of the elementary theory of numbers, i.e. rational numbers; the finding of the greatest common measure and the least common multiple, the theory of proportion applicable to rational numbers, propositions about numbers of different kinds, prime numbers, plane and solid numbers, squares, cubes and similar plane and solid numbers, series of numbers in continued proportion (i.e. in geometrical progression), the summation of a geometrical progression, the formation of 'perfect' numbers.

Book x contains a highly developed theory of irrationals, and is, in form, the most finished of all the Books. Euclid here owed much to the original investigations of Theodorus of Cyrene and Theaetetus of Athens.

Books XI, XII and XIII are on solid geometry, Book XI corresponding roughly in the character of its contents to Book I, Book XII to Books II and VI, and Book XIII to Book IV, in plane geometry. Book XIII shows how to construct all the five regular solids and to inscribe them respectively in given spheres, and ends with a proof that there can be no other regular solids. Book XIII is, in content, largely due to Theaetetus.

The Books purporting to be Books XIV and XV of the *Elements* are not by Euclid. Book XIV is by Hypsicles and is an interesting supplement to Book XIII. The main problems solved are the comparisons (1) of the sides (i.e. edges) of the cube and the icosahedron respectively inscribed in one and the same sphere, (2) of the surfaces and solid contents respectively of the dodecahedron and icosahedron inscribed in one and the same sphere. The results are those which Apollonius had established in a tract on the subject. Book XV is of much less interest, though it is also supplementary to the geometry of the five regular solids. One part of it at least was written in the sixth century A.D. by a pupil of Isidorus of Miletus, architect of the great church of St Sophia at Constantinople (about 532 A.D.).

THE ELEMENTS IN GREECE

THERE is no doubt that Euclid's *Elements* at once superseded all the earlier text-books, and took its place as the accepted authority on the subject. Archimedes, for instance, refers to Euclid by chapter and verse, as it were, and so do all later Greek geometers. It is true that—to say nothing of the attacks upon the *Elements* by those who, like the Epicureans, objected to the whole of mathematics— the methods of proof etc. were not to pass unquestioned for very long. Only a century later no less a person than the ' Great Geometer,' Apollonius of Perga, undertook the *rôle* of reformer. The suggestions of Apollonius seem to have been contained in a certain *General Treatise* (ἡ καθόλου πραγματεία), which apparently related to elementary geometry. In particular, Apollonius gave a new general definition of an angle, tried (vainly of course) to prove certain of the axioms, and gave alternative constructions for the problems of 1, 10, 11 and 23. These last seem to show that Apollonius wished to give a more *practical* turn to the beginnings of the subject; he seems in fact to have gone on the tack of combining theoretical with practical geometry which has found so much favour of late. So true is it that ' there is nothing new under the sun '; or, as Aristotle forcibly puts it, ' it is not once nor twice but times without number that the same views make their appearance in the world.' But the authority of Euclid remained unshaken. Proclus could say truly that in all the essentials for a text-book of Elements Euclid's system 'will be found superior to the rest'; indeed we hear of no rival to it in Greek geometry.

It was natural that so great a classic should become the subject of commentaries, more or less systematic, by later Greek mathematicians. We possess the commentary of Proclus on Book I; but the rest are lost except for the considerable number of extracts preserved in Proclus or in the Arabic.

1. The commentary of Heron of Alexandria (date still uncertain, the limits being 50 B.C. and the third century A.D.) was evidently rather elaborate. It is cited by Proclus, but more extensively by the Arabian commentator an-Nairīzī (died about 922). The last reference to Heron in an-Nairīzī is in the note to VIII, 27, which shows that Heron's commentary must have reached at least as far as that point. It contained a few general notes, but seems to have consisted, in the main, of what formed the ordinary commentator's stock-in-trade, dis-tinctions of cases, alternative proofs, with a few additions to or extensions of Euclid's propositions (notably in connexion with I, 47 and III, 20 re-spectively). It would appear that Heron was the first to adopt the quasi-algebraical method of proving the propositions of Book II (from the second proposition onwards) without a figure (except one line), as consequences of Prop. I.

2. Pappus (end of third century A.D.) wrote a commentary which was doubtless fairly complete. Proclus on Book I makes a few references to Pappus by name, and probably Pappus is drawn upon in other places where no name is mentioned. It is certain that Pappus wrote a commentary on Book X in two parts; fragments of this survive in an Arabic translation in a Paris MS. discovered by Woepcke. Lastly, Eutocius quotes a note from the commentary of Pappus on the *Elements*, the proper place of which is probably Book XII.

3. It is probable, though not certain, that Porphyry (about 232–304 A.D.), the Neo-Platonist, wrote a systematic commentary on part at least of the *Elements*. Proclus on Book I quotes some alternative proofs as given by Porphyry; they are of no particular importance.

4. Simplicius (fl. about 500 A.D.), the commentator on Aristotle, wrote a commentary on the beginning of the *Elements*, i.e. the definitions, postulates and axioms, which is preserved in an-Nairizi's commentary already mentioned. It is interesting for, among other things, an account of an attempt by a certain 'Aganis' to prove the Parallel-Postulate.

5. I have left till the last the commentary on Book I by Proclus (410–485 A.D.). The importance of this can hardly be over-estimated, since it is one of the two main sources of information on the history of Greek mathematics, the other being the

great *Collection* of Pappus. Proclus succeeded Syrianus as head of the Neo-Platonic school. He was an acute dialectician and pre-eminent among his contemporaries in the range of his learning. Although he was much more a philosopher than a mathematician, he was thoroughly versed both in mathematics and in astronomy.

The commentary seems to have been founded on lectures which Proclus gave to beginners in mathematics. But there are signs that it was revised and re-edited for a wider public, since he speaks in one place of ' those who shall come upon ' his work, and passages are included (e.g. about the cylindrical helix, the conchoid and the cissoid) which could hardly have been understood by beginners.

The importance of the commentary from a historical point of view lies in the fact that Proclus had access to a number of sources now lost. The following may be mentioned: Eudemus's *History of geometry* (the loss of which is one of the gravest that fate has inflicted on us), Geminus *On the theory of the mathematical sciences*, commentaries on Euclid by Heron, Pappus and probably Porphyry, a work on elementary geometry by Apollonius of Perga, Ptolemy *On the Parallel-Postulate*, books on *The Angle* by Eudemus and Syrianus respectively, Apollonius's tract *On the cochlias* (cylindrical helix), a whole book of Posidonius directed against the Epicurean Zeno, Carpus's *Astronomy*.

The most famous passage in the whole of the

commentary is the so-called 'Eudemian' summary of the history of geometry up to the immediate predecessors of Euclid, ending with the words, 'Those who have compiled histories bring the development of this science up to this point. Not much later than these is Euclid, who put together the Elements, collecting many of the theorems of Eudoxus, perfecting many others by Theaetetus, and bringing to irrefragable demonstration the things which had only been somewhat loosely proved by his predecessors.' It is certain that the summary up to this point must have been made up from scattered notices found in the great work of Eudemus; it is hardly possible that it can be textually quoted from Eudemus, and it seems more probable that it was taken from a compendium of Eudemus's history made by some later writer.

It is not certain whether Proclus continued his commentary beyond Book I. He intended to do so, but, at the time when the commentary on Book I was written, he had not begun to write on the other Books and was uncertain whether he would be able to do so; for he says at the end, 'For my part, if I should be able to discuss the other books in the same manner, I should give thanks to the gods; but, if other cares should draw me aside, I beg those who are attracted by the subject to complete the exposition of the other books as well, following the same method, and addressing themselves to the deeper and better defined questions involved.' As however the scholia which we possess show that their authors had the commentary on Book I in the same form as we have it, while they contain no trace of notes which Proclus promised in certain passages of the commentary on Book I, it is more than probable that his commentary extended no further.

CHANGES IN THE TEXT

IT would have been surprising if the text of a classic in such constant use and so much canvassed by a succession of commentators had not, in process of time, suffered a certain amount of alteration. A dividing line in the history of the text is marked by the new edition brought out by Theon of Alexandria in the fourth century A.D. It is well known that the title-page of Simson's edition of Euclid claims that in it 'the errors by which Theon or others have long ago vitiated these books are corrected and some of Euclid's demonstrations are restored '; and readers of Simson's notes will remember the familiar phrases in which, when anything in the text does not seem to him satisfactory, he says that the demonstration has been spoiled, or things have been interpolated or omitted, ' by Theon or some other unskilful editor.' Now most of the MSS. of the Greek text prove by their titles that they proceed from the recension of the *Elements* by Theon; they purport to be either ' from the edition of Theon ' or ' from the lectures of Theon ' (ἀπὸ συνουσιῶν τοῦ Θέωνος).

Theon himself says, in a passage of his Commentary on Ptolemy's *Syntaxis*, ' But that sectors in equal circles are to one another as the angles on which they stand has been proved by me in my edition of the Elements at the end of the sixth book.' Hence it is clear, not only that Theon edited the *Elements*, but that the second part of VI, 33 containing the proof of this proposition and found in nearly all the MSS. was added by him to the original.

The passage from Theon just quoted is the key to the whole question of Theon's changes in the text of the *Elements*; for when Peyrard found in the Vatican MS. 190 a text which contained neither the words from the titles of the other MSS. quoted above nor the interpolated second part of VI, 33, he was justified in concluding, as he did, that in the Vatican MS. (P) we have an edition more ancient than Theon's. It is clear too from a marginal note that the copyist of P or rather its archetype had before him the two recensions and systematically gave the preference to the earlier one. We are thus more fortunate than Simson in that we can form our judgment of Theon's recension by the actual documentary evidence furnished by a comparison of the Vatican MS. P with the MSS. which we may, after Heiberg, conveniently call the Theonine MSS.

We have naturally to look elsewhere for evidence of interpolations in the text before Theon's time. Besides internal evidence, Heiberg uses for this purpose (1) some valuable fragments of ancient papyri, and (2) the fresh evidence furnished by an-Nairīzī's commentary with its rich quotations from Heron, which in several cases enable us to detect differences between our text and the text as Heron had it, and to identify some interpolations which actually found their way into the text from Heron's commentary itself (notably the alternative proof of III, 10 and the substantive proposition III, 12).

The interpolations made before Theon's time may be thus classified:

(1) Alternative proofs and additional cases, the former often betraying themselves by introductory words such as ' We can prove this *more readily* ' or '*more shortly* thus.'

(2) Lemmas, of which there are a considerable number, especially in Books X–XIII.

(3) Porisms (or corollaries). Sometimes parts only of porisms are interpolated, and sometimes the interpolation consists of a sort of proof (as if the interpolator thought the porism not obvious enough).

(4) Scholia, sometimes betrayed by the use of the words καλεῖ and ἐκάλεσε, 'he calls' and 'he called' (sc. Euclid).

(5) Some whole propositions, e.g. I, 41, III, 12 (due to Heron), and at least one proposition (117) of Book X containing the traditional proof (referred to by Aristotle and doubtless Pythagorean) of the fact that the diagonal of a square is incommensurable with the side.

(6) A number of definitions (e.g. the definition of a 'segment of a circle' in Book I and that of 'reciprocal figures,' VI, Def. 2) and axioms (most of those after *Common Notions* 1–3).

It would appear that the text was most spoiled by interpolations in the third century A.D., since Sextus Empiricus (about 180 A.D.) had a correct text, while Iamblichus (early in fourth century) had an interpolated one.

With regard to the changes introduced by Theon the following notes may be added. Although Theon took some trouble, in cases where his sources differed, to follow older MSS., his object was less to discover and restore from MSS. the actual words written by Euclid than to remove difficulties that might be felt by learners in studying the book. We have (1) alterations made by Theon where he found, or thought he found, blots in the original, (2) emendations intended to improve the form or diction of Euclid; but he seems to have given the most attention to (3) additions designed to supplement or explain Euclid. Under the head of additions come the second part of VI, 33 relating to *sectors*, a second case to VI, 27, some porisms and some alternative proofs. Under the head of explanations come (*a*) repetitions of the argument where Euclid had said 'For the same reason,' (*b*) intermediate steps supplied where Euclid's argument seemed too rapid and not easy enough to follow, and (*c*) additions of a word or two for the sake of clearness and consistency. On the other hand he made (*d*) some omissions which were designed, like the additions, to reduce the language to a standard form. In particular, Theon seems to be responsible for the frequent omission of the words ὅπερ ἔδει δεῖξαι (or ποιῆσαι), Q.E.D. (or F.), at the end of propositions. A curious point is that, when Euclid added a porism (corollary) at the end of a proposition, it was his habit to omit the words ὅπερ ἔδει δεῖξαι (ποιῆσαι) at the end of the proposition proper and to insert them after the porism, as if he regarded the latter as being actually a part of the proposition. In the Theonine MSS., on the other hand, the words are sometimes omitted and sometimes inserted at the end of the proposition instead of at the end of the porism.

There is no doubt that Theon's edition was approved by the persons for whom it was written; and it was almost exclusively used by later Greeks, with the result, as above mentioned, that the more ancient text is preserved to us in only one complete MS.

LATER HISTORY OF THE ELEMENTS

WE have next to consider the fate of the *Elements* in countries outside Greece.

Italy was not a favourable soil. Cicero is the first Latin author to mention Euclid; but he also tells us that, while geometry was held in high honour among the Greeks, the Romans limited its scope by having regard only to its utility for measurements and calculations. The Roman *agrimensores* were satisfied with the minimum of theoretical geometry: cf. Balbus *de mensuris*, where some of the definitions of Book I are given. Again, the extracts from the *Elements* in a fragment attributed to Censorinus (fl. 238 A.D.) are confined to the definitions, postulates and axioms. By degrees, however, the *Elements* passed into the curriculum of a liberal education. Martianus Capella (*circa* 470 A.D.) describes the effect of the mention of the enunciation of the problem how to construct an equilateral triangle on a given straight line in a company of philosophers, who forthwith broke out into encomiums on Euclid. But the *Elements* were probably at that time read in Greek.

Magnus Aurelius Cassiodorius (born about 475 A.D.) says that Euclid was translated into the Latin language by Boëtius. A letter of Theodoric to Boëtius also contains the words ' for in your translations...Nicomachus the arithmetician and Euclid the geometer are heard in the Ausonian tongue.' The translation can hardly be represented by the so-called *Geometry* of Boëtius which has come down to us, for this by no means constitutes a translation of Euclid. Even the version of it in two Books which was edited by Friedlein is not genuine, but appears to have been put together in the eleventh century from various sources. It begins with the definitions of Book I; then come the postulates (five) and the axioms (three), followed by the *enunciations* only (without proofs) of the propositions of Book I, of ten propositions of Book II and of some propositions from Books III and IV; these are followed by a literal translation of the *proofs* of Euclid I, 1-3, which shows that the Pseudo-Boëtius had before him a Latin translation from which he extracted these three proofs.

The *Elements* next passed to Arabia. The Caliph al-Manṣūr (754–775), as the result of a mission to the Byzantine Emperor, obtained from him a copy of Euclid among other Greek books, and again the Caliph al-Ma'mūn (813–833) acquired MSS. of Euclid among others from the Byzantines. The first translation of the *Elements* into Arabic was made by al-Ḥajjāj b. Yūsuf b. Maṭar, who seems to have translated them twice, first for Hārūn ar-Rashīd (786–809) and then for al-Ma'mūn. Six Books of the second version survive in a Leiden MS. (Cod. Leidensis 399, 1) which is being published, with a Latin translation, by Besthorn and Heiberg. The object of al-Ḥajjāj seems to have been less to give a faithful reflection of the original than to write a useful and convenient mathematical text-book. This may be the reason why a fresh translation was undertaken by Isḥāq b. Ḥunain (the son of the more famous translator Ḥunain b. Isḥāq al-'Ibādī). Isḥāq died in 910, but already in his lifetime the translation was revised by Thābit b. Qurra (d. 901). The translation by Isḥāq was made direct from the

Greek, and Thābit consulted Greek MSS. in making the revision. The original version by Isḥāq does not seem to have survived, but the Isḥāq-Thābit version is extant in two MSS. at Oxford containing Books I–XIII and one at Copenhagen containing Books V–X. Thābit's version, which is said to be a model of a translation of a mathematical text, was translated into Latin by Gherard of Cremona (1114–1187), and MSS. were discovered in 1904 by Axel Anthon Björnbo which seem to contain this translation. It was made from a MS. containing a revised and critical edition of Thābit's version; the translator also compared the Thābit version with other Arabic versions, and apparently with some translation from the Greek. The third form of the Arabian Euclid actually accessible is the edition of Naṣīraddīn aṭ-Ṭūsī (1201–1274), which exists in two forms, a longer and a shorter. Aṭ-Ṭūsī's work was not a translation of Euclid's text, but a re-written Euclid based on the older Arabic translations.

In 747 the Arabs who crossed from Africa to Spain established themselves there under a separate caliphate. In due course Arabian learning also found a home in the West and, under the most enlightened of the caliphs, schools and libraries were established at such places as Seville, Cordova and Granada. We hear of a palace at Cordova containing a library of 600,000 volumes, though the figure may be an exaggeration. It was in Spain that the author of the first extant translation of the *Elements* into

Latin, Athelhard of Bath, got his copy of the Arabic version. Athelhard is said to have studied at Tours and Laon and to have travelled in Greece, Asia Minor and Egypt as well as Spain; he went to Cordova disguised as a Mohammedan student. His translation of the *Elements* was made about 1120 A.D. (Athelhard was however not the first to translate the *Elements* into Latin; there are traces of an earlier translation apparently made from the Greek, which must have been in existence before the eleventh century. That some sort of translation, or at least fragments of one, were available before Athelhard's time even in England is indicated by the old English verses:

The clerk Euclide on this wyse hit fonde
Thys craft of gemetry yn Egypte londe.
Yn Egypte he tawghte hyt ful wyde,
In dyvers londe on every syde.
Mony erys afterwarde y understonde
Yer that the craft com ynto thys londe.
Thys craft com into England, as y yow say,
Yn tyme of good Kyng Adelstone's day,

which would put the introduction of Euclid into England as far back as 924–940 A.D.)

Closely related to the translation by Athelhard is that of Johannes Campanus (Campano), who is mentioned by Roger Bacon (1214–1294) as a prominent mathematician of his time and who was chaplain to Pope Urban IV (1261–1281). It would appear that Campanus used Athelhard's translation (the definitions, postulates and axioms, and the 364 enunciations are word for word identical in the two translations), but developed the proofs by means of another redaction of the Arabian Euclid. Campanus's translation is the more complete, and his arrangement is different from Athelhard's in that in Athelhard the proofs regularly precede the enunciations, whereas Campanus follows the usual order. It has been suggested that Athelhard's translation may have been gradually altered by successive copyists and commentators who had Arabic originals before them until it took the form in which Campanus gave it to the world.

The first-fruits of the discovery of the Greek masterpieces in the Arabic translations were soon seen in the works of a very brilliant mathematician, Leonardo of Pisa, called Fibonacci (about 1175–1250). Leonardo assimilated all that the Arabs could tell him about arithmetic and algebra (this is clear from his *Liber abbaci*), and similarly in his *Practica geometriae* (1220) he gives in substance all that he could learn from the *Elements* of Euclid (whom he often mentions by name), and from Archimedes's works about the measurement of plane figures bounded by straight lines, solid figures bounded by planes, the circle and the sphere, besides the elements of trigonometry as given by Ptolemy. What is remarkable about Leonardo is the complete grasp which he shows over the whole of the subject-matter. There is nothing about him of the diffidence of the novice who adheres painfully to the text and form of the work drawn upon, hugging the shore as it were. On the contrary he launches out boldly, and alters the form of exposition at his pleasure and with the utmost certainty, yet abating nothing from absolute rigour of proof and often adding clever developments of his own. It is extraordinary that, after such a send-off, geometry should have languished completely for two centuries and more. Nothing in the least comparable to Leonardo's work appeared in the fourteenth and fifteenth centuries.

It was not till after the invention of printing that the period of stagnation came to an end. In 1482 appeared the first printed edition of the *Elements*. This fine book was printed at Venice by Erhard Ratdolt, and contained Campanus's translation. The book has a margin of $2\frac{1}{2}$ inches, and in this margin are placed the figures of the propositions. Ratdolt claims that, while up to that time no one had succeeded in printing diagrams, he had after much labour discovered a method by which figures could be produced as easily as letters. How eagerly the opportunity of spreading geometrical knowledge was seized upon is shown by the number of editions which followed in the next few years. Even the year 1482 saw two forms of the book, though they only differ in the first sheet. Others came out in 1486 and 1491 respectively.

In 1505 Bartolomeo Zamberti (Zambertus) brought out at Venice the first translation from the

Greek text of the whole of the *Elements*. Zamberti quite unfairly attacked Campanus as 'that most barbarous translator' because he had used such terms as 'helmuain' and 'helmuariphe'; Zamberti does not seem to have had the penetration to see that Campanus was translating, not from the Greek, but from the Arabic.

The *editio princeps* of the Greek text appeared in 1533 at Basel. The editor was Simon Grynaeus the elder (d. 1541), and the preface is addressed to a notable Englishman, Cuthbert Tonstall (1474–1559), the author of a book on arithmetic, *De arte supputandi libri quatuor*. Unfortunately the two MSS. used by Grynaeus are among the worst, so that his text could not claim to be authoritative. Yet it remained for a long time the source and foundation of later editions of the Greek text.

1570 saw the first English translation of the whole of the *Elements*. The translator was Sir Henry Billingsley, who had a varied career. Admitted Lady Margaret scholar of St John's College, Cambridge, in 1551, he is said also to have studied at Oxford, though he did not take a degree at either University. He was afterwards apprenticed to a London haberdasher and rapidly became a wealthy merchant. Sheriff of London in 1584, he was elected Lord Mayor on 31 December, 1596, on the death, during his year of office, of Sir Thomas Skinner. From 1589 he was one of the Queen's four 'customers,' or farmers of customs, of the port of

London. In 1591 he founded three scholarships for poor students at St John's College, Cambridge. Billingsley's translation is a magnificent book of 928 folio pages (excluding the long preface of John Dee). Billingsley was evidently a competent Greek scholar. This appears from marginal notes in the copy of the Greek text used by him (now in Princeton College, U.S.A.), which contain conjectural emendations of the text as well as comments on the translations by Athelhard and Campanus from the Arabic. The commentary in Billingsley's edition includes all the most important notes that had ever been written from the time of the Greek commentators onwards. The elaborateness of the book may be illustrated by the fact that the diagrams of Book XI are nearly all duplicated, one figure being that of Euclid, and the other an arrangement of little pieces of paper (triangular, rectangular, etc.) pasted at the edges on to the pages of the book so as to admit of being turned up and made to show the real form of the solid figure depicted.

The most important Latin translation is that of Commandinus of Urbino (1509–1575), which was the basis of most of the later translations up to the beginning of the nineteenth century, including that of Simson and therefore of those editions also, numerous in England, which give Euclid 'chiefly after the text of Simson.' The date of publication was 1572. Commandinus seems to have used, in addition to the Basel *editio princeps*, some Greek MS., so far not identified. He not only followed the original Greek more closely than his predecessors, but added to his translation some ancient scholia as well as good notes of his own.

The Latin version by Clavius (Christoph Klau), which appeared in 1574, was not a translation but a re-written version, with proofs compressed or added to whenever the editor thought they could be made clearer. The book contains a very extensive and valuable collection of notes. This version also went through several editions.

In 1654 appeared André Tacquet's *Elementa geometriae planae et solidae* containing the eight geometrical Books arranged for general use in schools. This edition had a great success; many times re-issued up to the end of the eighteenth century, it held its own as the standard text-book in use on the continent until superseded by Legendre.

Barrow's *Euclidis Elementorum libri xv breviter demonstrati* (1655) compressed the whole of the *Elements* into a very small compass by abbreviating the proofs and using a large quantity of symbols (which, Barrow says, were generally Oughtred's). In the preface to the edition of 1659 he says that he would not have written it but for the fact that Tacquet gave only eight Books of Euclid. There were several editions of the book up to 1732 (those of 1660 and 1732 and one or two others are in English). It remained a standard text-book till the beginning of the eighteenth century.

1703 is the date of the great Oxford edition of the Greek text by David Gregory, which, until the appearance of Heiberg and Menge's new text, was still the only edition of the complete works of Euclid. In the Latin translation accompanying the Greek text Gregory mainly followed Commandinus; the text itself was based on the *editio princeps* except in certain special passages where, with the help of John Hudson, Bodley's librarian, some of the Greek MSS. bequeathed by Savile to the University were consulted.

In 1703 William Whiston, Lucasian Professor at Cambridge, brought out an edition of Tacquet's Latin version of the eight geometrical Books. A second edition followed in 1710 and a third (in English and abridged) in 1714. The book went through several more editions and was the standard English text-book on elementary geometry until superseded by Simson's.

In 1708 appeared an English translation of Books I–VI, XI and XII from Commandinus's version by Dr John Keill, Savilian Professor of Geometry at Oxford. The book was revised and corrected in 1723 by Samuel Cunn. From the number of editions (1749 saw the eighth and 1782 the twelfth) we may conclude that it had considerable vogue.

Simson's first edition, in Latin and in English, appeared in 1756 at Glasgow; it contained the first six Books with the eleventh and twelfth. The title-page claims that ' in this edition the errors by which

Theon or others have long ago vitiated these Books
are corrected and some of Euclid's Demonstrations
are restored.' It is true that many of the faults were
not due to Theon or even to others, and that, in
preferring certain alternative proofs and making
certain alterations in the text, Simson did not
always show sound judgment. But on the other
hand there are many cases in which he found real
blots and showed how to remove them. This remark applies especially to Book v (on proportion)
where Simson's acuteness both in criticism and
emendation is seen at its best. The notes attached
at the end of the work describe the changes made
by Simson as compared with the Greek text and the
reasons for the changes, and they contain, in
addition, many valuable explanations and observations. Simson's version fully deserved to become
a standard text-book; and some thirty editions
attest its actual success.

The translation by James Williamson of the whole
thirteen Books (vol. i, Oxford 1781, vol. ii, London
1788) was, up to 1908, the last which reproduced
Euclid word for word, though from the imperfect
text then in use.

Peyrard's Greek text published in three volumes
between 1814 and 1818 represents the first approach
to a better text in so far as it adopted or recorded
the readings of the Vatican MS. 190 (P). It was
accompanied by translations into Latin and French.

A most valuable edition of Books i–vi in Greek and

Latin is that of J. G. Camerer (and C. F. Hauber) in two volumes published at Berlin 1824–25. The Greek text is mainly based on Peyrard, although the Basel and Oxford editions were also used. The collection of notes is probably the most complete that exists, and is well-nigh inexhaustible.

E. F. August's Greek text (1826–29) containing Books I–XIII further improved on Peyrard's; but all earlier editions of Euclid in Greek are now superseded by the definitive text edited by Heiberg and Menge in the Teubner series. The *Elements* (edited by Heiberg) with prolegomena, critical notes, scholia, etc., appeared as vols. I–V in the years from 1883 to 1888.

A new translation, from Heiberg's text, of the whole of the thirteen Books of the *Elements*, with introduction and commentary, is now available (T. L. Heath, Cambridge, 1908, three volumes).

EUCLID IN EDUCATION

In adding a few notes on the part played by Euclid in education we need only begin with the century following that in which the first Latin translations by Athelhard and Gherard of Cremona were made. Some idea of the general neglect of geometry in the thirteenth century may be gathered from remarks of Roger Bacon (1214–1294). He tells us that in his time the University of Paris paid little attention to mathematics, and he has nothing good to say of those who taught there. At Oxford he says that there were few students who (unless driven to it by the rod) cared to do more than three or four propositions of Euclid, and it was in consequence of this that the fifth proposition was called '*Elefuga*, fuga miserorum,' or, as we might say, 'escape from troubling' or 'end of troubling' (the play on the words ἔλεος and *elementum* can hardly be expressed in English). In the revised regulations of the University of Paris in 1366 it was laid down that candidates for the Licentiate must have heard lectures on some mathematical books (unspecified); and similar regulations are said to have been made a year or two later at Oxford and Cambridge. The preface to *Demonstrations on the first six Books of the Elements* by Oronce Fine (1536) says that at that time the University of Paris required from all who aspired to the laurels of philosophy the most binding form of oath that they had attended lectures on the first six Books of Euclid; there was no examination-test. Similarly at the University of Prag in 1367 the regulations prescribed lectures on six Books; at Vienna in 1389 candidates must have attended lectures, on one Book for the bachelor's degree, and on five Books for the Licentiate; at Cologne in 1398 the corresponding requirement for the Licentiate was limited to three Books; and so on. At Oxford between 1449 and 1463 the only mathematical subjects read were Ptolemy's astronomy and the first two Books of Euclid; and the conditions were no doubt similar at Cambridge. The Edwardian statutes of the University of Cambridge laid it down that freshmen should begin with mathematics as being the best foundation for a liberal education; and Euclid is mentioned as the text-book of geometry. In 1550 John Dee, Fellow of St John's College, Cambridge, was lecturing in Paris, in English, on Euclidean geometry to large audiences and—contrary to the prevailing practice—gratuitously. Sir Henry Savile (1549–1622) who founded the Savilian Professorships of Geometry and Astronomy in 1619, gave lectures on Euclid, also free, in 1620. There were thirteen of these lectures, and they were published in 1621. They did not extend beyond 1, 8, but they are valuable because they grapple with the difficulties connected with the preliminary notions, the definitions, postulates, etc., and the tacit assumptions. Briggs, the first Savilian Professor of Geometry, took up the subject where Savile left off. But it was during the seventy years or so from 1660 to 1730 when Wallis and Halley were Professors at Oxford, and Barrow and Newton at Cambridge, that the study of Greek geometry was at its height in England. After Newton's time attention would of course be mainly directed to the new analytical methods in mathematics and natural philosophy; and the Greeks (except Euclid) fell into the background. But it is rather surprising that as late as 1799 'a knowledge of the first two Books of Euclid, algebra to simple and quadratic equations, and the early chapters of Paley's *Evidences of Christianity* was still considered sufficient to secure a position in the senior optimes' (W. W. Rouse Ball, *History of Mathematics at Cambridge*, p. 198). It is conjectured that Euclid became a *school* text-book about the middle of the eighteenth century when boys came to stay longer at school and therefore required to have some preliminary knowledge of mathematics before going up to the University (Gow, *History of Greek Mathematics*, p. 208).

GREEK TEXT OF BOOK I

ΣΤΟΙΧΕΙΩΝ α΄.

Ὅροι.

α΄. ΣΗΜΕΙΟΝ ἐστιν, οὗ μέρος οὐθέν.

β΄. Γραμμὴ δὲ μῆκος ἀπλατές.

γ΄. Γραμμῆς δὲ πέρατα σημεῖα.

δ΄. Εὐθεῖα γραμμή ἐστιν, ἥτις ἐξ ἴσου τοῖς ἐφ᾽ ἑαυτῆς σημείοις κεῖται.

ε΄. Ἐπιφάνεια δέ ἐστιν, ὃ μῆκος καὶ πλάτος μόνον ἔχει.

ϛ΄. Ἐπιφανείας δὲ πέρατα γραμμαί.

ζ΄. Ἐπίπεδος ἐπιφάνειά ἐστιν, ἥτις ἐξ ἴσου ταῖς ἐφ᾽ ἑαυτῆς εὐθείαις κεῖται.

η΄. Ἐπίπεδος δὲ γωνία ἐστὶν ἡ ἐν ἐπιπέδῳ δύο γραμμῶν ἁπτομένων ἀλλήλων καὶ μὴ ἐπ᾽ εὐθείας κειμένων πρὸς ἀλλήλας τῶν γραμμῶν κλίσις.

θ΄. Ὅταν δὲ αἱ περιέχουσαι τὴν γωνίαν γραμμαὶ εὐθεῖαι ὦσιν, εὐθύγραμμος καλεῖται ἡ γωνία.

ι΄. Ὅταν δὲ εὐθεῖα ἐπ᾽ εὐθεῖαν σταθεῖσα τὰς ἐφεξῆς γωνίας ἴσας ἀλλήλαις ποιῇ, ὀρθὴ ἑκατέρα τῶν ἴσων γωνιῶν ἐστί, καὶ ἡ ἐφεστηκυῖα εὐθεῖα κάθετος καλεῖται, ἐφ᾽ ἣν ἐφέστηκεν.

ια΄. Ἀμβλεῖα γωνία ἐστὶν ἡ μείζων ὀρθῆς.

ιβ΄. Ὀξεῖα δὲ ἡ ἐλάσσων ὀρθῆς.

ιγ΄. Ὅρος ἐστίν, ὃ τινός ἐστι πέρας.

ιδ΄. Σχῆμά ἐστι τὸ ὑπό τινος ἢ τινων ὅρων περιεχόμενον.

ιε΄. Κύκλος ἐστὶ σχῆμα ἐπίπεδον ὑπὸ μιᾶς γραμμῆς περιεχόμενον [ἣ καλεῖται περιφέρεια], πρὸς ἣν ἀφ᾽ ἑνὸς σημείου τῶν ἐντὸς τοῦ σχήματος κειμένων πᾶσαι αἱ προσπίπτουσαι εὐθεῖαι [πρὸς τὴν τοῦ κύκλου περιφέρειαν] ἴσαι ἀλλήλαις εἰσίν.

ιϛ΄. Κέντρον δὲ τοῦ κύκλου τὸ σημεῖον καλεῖται.

ιζ΄. Διάμετρος δὲ τοῦ κύκλου ἐστὶν εὐθεῖά τις διὰ τοῦ κέντρου ἠγμένη καὶ περατουμένη ἐφ᾽ ἑκάτερα τὰ μέρη ὑπὸ τῆς τοῦ κύκλου περιφερείας, ἥτις καὶ δίχα τέμνει τὸν κύκλον.

ιη΄. Ἡμικύκλιον δέ ἐστι τὸ περιεχόμενον σχῆμα ὑπό τε τῆς διαμέτρου καὶ τῆς ἀπολαμβανομένης ὑπ᾽ αὐτῆς περιφερείας. κέντρον δὲ τοῦ ἡμικυκλίου τὸ αὐτό, ὃ καὶ τοῦ κύκλου ἐστίν.

ιθ΄. Σχήματα εὐθύγραμμά ἐστι τὰ ὑπὸ εὐθειῶν περιεχόμενα, τρίπλευρα μὲν τὰ ὑπὸ τριῶν, τετράπλευρα δὲ τὰ ὑπὸ τεσσάρων, πολύπλευρα δὲ τὰ ὑπὸ πλειόνων ἢ τεσσάρων εὐθειῶν περιεχόμενα.

κ΄. Τῶν δὲ τριπλεύρων σχημάτων ἰσόπλευρον μὲν τρίγωνόν ἐστι τὸ τὰς τρεῖς ἴσας ἔχον πλευράς, ἰσοσκελὲς δὲ τὸ τὰς δύο μόνας ἴσας ἔχον πλευράς, σκαληνὸν δὲ τὸ τὰς τρεῖς ἀνίσους ἔχον πλευράς.

κα΄. Ἔτι δὲ τῶν τριπλεύρων σχημάτων ὀρθογώνιον μὲν τρίγωνόν ἐστι τὸ ἔχον ὀρθὴν γωνίαν, ἀμβλυγώνιον δὲ τὸ ἔχον ἀμβλεῖαν γωνίαν, ὀξυγώνιον δὲ τὸ τὰς τρεῖς ὀξείας ἔχον γωνίας.

κβ΄. Τῶν δὲ τετραπλεύρων σχημάτων τετράγωνον μέν ἐστιν, ὃ ἰσόπλευρόν τέ ἐστι καὶ ὀρθογώνιον,

ἑτερόμηκες δέ, ὃ ὀρθογώνιον μέν, οὐκ ἰσόπλευρον
δέ, ῥόμβος δέ, ὃ ἰσόπλευρον μέν, οὐκ ὀρθογώνιον
δέ, ῥομβοειδὲς δὲ τὸ τὰς ἀπεναντίον πλευράς τε
καὶ γωνίας ἴσας ἀλλήλαις ἔχον, ὃ οὔτε ἰσόπλευρόν ἐστιν οὔτε ὀρθογώνιον· τὰ δὲ παρὰ
ταῦτα τετράπλευρα τραπέζια καλείσθω.

κγ΄. Παράλληλοί εἰσιν εὐθεῖαι, αἵτινες ἐν τῷ αὐτῷ
ἐπιπέδῳ οὖσαι καὶ ἐκβαλλόμεναι εἰς ἄπειρον
ἐφ᾽ ἑκάτερα τὰ μέρη ἐπὶ μηδέτερα συμπίπτουσιν
ἀλλήλαις.

Αἰτήματα.

α΄. Ἡιτήσθω ἀπὸ παντὸς σημείου ἐπὶ πᾶν σημεῖον
εὐθεῖαν γραμμὴν ἀγαγεῖν.

β΄. Καὶ πεπερασμένην εὐθεῖαν κατὰ τὸ συνεχὲς ἐπ᾽
εὐθείας ἐκβαλεῖν.

γ΄. Καὶ παντὶ κέντρῳ καὶ διαστήματι κύκλον γράφεσθαι.

δ΄. Καὶ πάσας τὰς ὀρθὰς γωνίας ἴσας ἀλλήλαις
εἶναι.

ε΄. Καὶ ἐὰν εἰς δύο εὐθείας εὐθεῖα ἐμπίπτουσα τὰς
ἐντὸς καὶ ἐπὶ τὰ αὐτὰ μέρη γωνίας δύο ὀρθῶν
ἐλάσσονας ποιῇ, ἐκβαλλομένας τὰς δύο εὐθείας
ἐπ᾽ ἄπειρον συμπίπτειν, ἐφ᾽ ἃ μέρη εἰσὶν αἱ τῶν
δύο ὀρθῶν ἐλάσσονες.

Κοιναὶ ἔννοιαι.

α'. Τὰ τῷ αὐτῷ ἴσα καὶ ἀλλήλοις ἐστὶν ἴσα.

β'. Καὶ ἐὰν ἴσοις ἴσα προστεθῇ, τὰ ὅλα ἐστὶν ἴσα.

γ'. Καὶ ἐὰν ἀπὸ ἴσων ἴσα ἀφαιρεθῇ, τὰ καταλειπόμενά ἐστιν ἴσα.

ζ'. Καὶ τὰ ἐφαρμόζοντα ἐπ' ἄλληλα ἴσα ἀλλήλοις ἐστίν.

η'. Καὶ τὸ ὅλον τοῦ μέρους μεῖζόν [ἐστιν].

α'.

Ἐπὶ τῆς δοθείσης εὐθείας πεπερασμένης τρίγωνον ἰσόπλευρον συστήσασθαι.

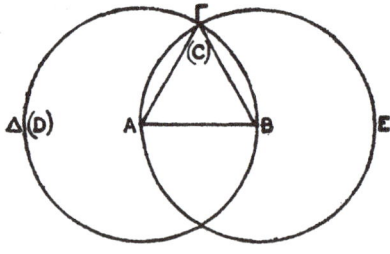

ΕΣΤΩ ἡ δοθεῖσα εὐθεῖα πεπερασμένη ἡ ΑΒ.

Δεῖ δὴ ἐπὶ τῆς ΑΒ εὐθείας τρίγωνον ἰσό-
5 πλευρον συστήσασθαι.

Κέντρῳ μὲν τῷ Α διαστήματι δὲ τῷ ΑΒ κύκλος
γεγράφθω ὁ ΒΓΔ, καὶ πάλιν κέντρῳ μὲν τῷ Β διαστήματι δὲ τῷ ΒΑ κύκλος γεγράφθω ὁ ΑΓΕ, καὶ
ἀπὸ τοῦ Γ σημείου, καθ᾽ ὃ τέμνουσιν ἀλλήλους οἱ
κύκλοι, ἐπὶ τὰ Α, Β σημεῖα ἐπεζεύχθωσαν εὐθεῖαι 10
αἱ ΓΑ, ΓΒ.

Καὶ ἐπεὶ τὸ Α σημεῖον κέντρον ἐστὶ τοῦ ΓΔΒ
κύκλου, ἴση ἐστὶν ἡ ΑΓ τῇ ΑΒ· πάλιν, ἐπεὶ τὸ Β
σημεῖον κέντρον ἐστὶ τοῦ ΓΑΕ κύκλου, ἴση ἐστὶν ἡ
ΒΓ τῇ ΒΑ. ἐδείχθη δὲ καὶ ἡ ΓΑ τῇ ΑΒ ἴση· 15
ἑκατέρα ἄρα τῶν ΓΑ, ΓΒ τῇ ΑΒ ἐστιν ἴση. τὰ δὲ
τῷ αὐτῷ ἴσα καὶ ἀλλήλοις ἐστὶν ἴσα· καὶ ἡ ΓΑ ἄρα
τῇ ΓΒ ἐστιν ἴση· αἱ τρεῖς ἄρα αἱ ΓΑ, ΑΒ, ΒΓ ἴσαι
ἀλλήλαις εἰσίν.

ἰσόπλευρον ἄρα ἐστὶ τὸ ΑΒΓ τρίγωνον. καὶ συν- 20
έσταται ἐπὶ τῆς δοθείσης εὐθείας πεπερασμένης τῆς
ΑΒ.

[Ἐπὶ τῆς δοθείσης ἄρα εὐθείας πεπερασμένης
τρίγωνον ἰσόπλευρον συνέσταται] ὅπερ ἔδει ποιῆσαι.

β΄.

Πρὸς τῷ δοθέντι σημείῳ τῇ δοθείσῃ εὐθείᾳ ἴσην
εὐθεῖαν θέσθαι.

ΕΣΤΩ τὸ μὲν δοθὲν σημεῖον τὸ Α, ἡ δὲ δοθεῖσα
εὐθεῖα ἡ ΒΓ· δεῖ δὴ πρὸς τῷ Α σημείῳ τῇ
δοθείσῃ εὐθείᾳ τῇ ΒΓ ἴσην εὐθεῖαν θέσθαι. 5

Ἐπεζεύχθω γὰρ ἀπὸ τοῦ Α σημείου ἐπὶ τὸ Β
σημεῖον εὐθεῖα ἡ ΑΒ, καὶ συνεστάτω ἐπ᾽ αὐτῆς

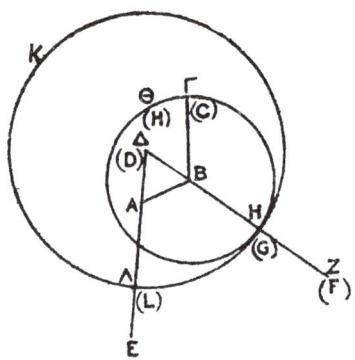

τρίγωνον ἰσόπλευρον τὸ ΔΑΒ, καὶ ἐκβεβλήσθωσαν
ἐπ᾽ εὐθείας ταῖς ΔΑ, ΔΒ εὐθεῖαι αἱ ΑΕ, ΒΖ, καὶ
10 κέντρῳ μὲν τῷ Β διαστήματι δὲ τῷ ΒΓ κύκλος
γεγράφθω ὁ ΓΗΘ, καὶ πάλιν κέντρῳ τῷ Δ καὶ
διαστήματι τῷ ΔΗ κύκλος γεγράφθω ὁ ΗΚΛ.
Ἐπεὶ οὖν τὸ Β σημεῖον κέντρον ἐστὶ τοῦ ΓΗΘ
<κύκλου>, ἴση ἐστὶν ἡ ΒΓ τῇ ΒΗ. πάλιν, ἐπεὶ τὸ Δ
15 σημεῖον κέντρον ἐστὶ τοῦ ΗΚΛ κύκλου, ἴση ἐστὶν ἡ
ΔΛ τῇ ΔΗ, ὧν ἡ ΔΑ τῇ ΔΒ ἴση ἐστίν. λοιπὴ ἄρα
ἡ ΑΛ λοιπῇ τῇ ΒΗ ἐστιν ἴση. ἐδείχθη δὲ καὶ ἡ ΒΓ
τῇ ΒΗ ἴση· ἑκατέρα ἄρα τῶν ΑΛ, ΒΓ τῇ ΒΗ ἐστιν
ἴση. τὰ δὲ τῷ αὐτῷ ἴσα καὶ ἀλλήλοις ἐστὶν ἴσα·
20 καὶ ἡ ΑΛ ἄρα τῇ ΒΓ ἐστιν ἴση.

Πρὸς ἄρα τῷ δοθέντι σημείῳ τῷ Α τῇ δοθείσῃ
εὐθείᾳ τῇ ΒΓ ἴση εὐθεῖα κεῖται ἡ ΑΛ· ὅπερ ἔδει
ποιῆσαι.

γ΄.

Δύο δοθεισῶν εὐθειῶν ἀνίσων ἀπὸ τῆς μείζονος
τῇ ἐλάσσονι ἴσην εὐθεῖαν ἀφελεῖν.

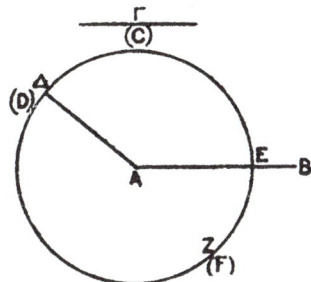

ΣΤΩΣΑΝ αἱ δοθεῖσαι δύο εὐθεῖαι ἄνισοι αἱ ΑΒ,
Γ, ὧν μείζων ἔστω ἡ ΑΒ· δεῖ δὴ ἀπὸ τῆς
μείζονος τῆς ΑΒ τῇ ἐλάσσονι τῇ Γ ἴσην εὐθεῖαν 5
ἀφελεῖν.

Κείσθω πρὸς τῷ Α σημείῳ τῇ Γ εὐθείᾳ ἴση ἡ
ΑΔ· καὶ κέντρῳ μὲν τῷ Α διαστήματι δὲ τῷ ΑΔ
κύκλος γεγράφθω ὁ ΔΕΖ.

Καὶ ἐπεὶ τὸ Α σημεῖον κέντρον ἐστὶ τοῦ ΔΕΖ 10
κύκλου, ἴση ἐστὶν ἡ ΑΕ τῇ ΑΔ· ἀλλὰ καὶ ἡ Γ τῇ
ΑΔ ἐστιν ἴση. ἑκατέρα ἄρα τῶν ΑΕ, Γ τῇ ΑΔ
ἐστιν ἴση· ὥστε καὶ ἡ ΑΕ τῇ Γ ἐστιν ἴση.

Δύο ἄρα δοθεισῶν εὐθειῶν ἀνίσων τῶν ΑΒ, Γ
ἀπὸ τῆς μείζονος τῆς ΑΒ τῇ ἐλάσσονι τῇ Γ ἴση 15
ἀφῄρηται ἡ ΑΕ· ὅπερ ἔδει ποιῆσαι.

δ΄.

Ἐὰν δύο τρίγωνα τὰς δύο πλευρὰς [ταῖς] δυσὶ
πλευραῖς ἴσας ἔχῃ ἑκατέραν ἑκατέρᾳ καὶ τὴν
γωνίαν τῇ γωνίᾳ ἴσην ἔχῃ τὴν ὑπὸ τῶν ἴσων
εὐθειῶν περιεχομένην, καὶ τὴν βάσιν τῇ βάσει
5   ἴσην ἕξει, καὶ τὸ τρίγωνον τῷ τριγώνῳ ἴσον
ἔσται, καὶ αἱ λοιπαὶ γωνίαι ταῖς λοιπαῖς
γωνίαις ἴσαι ἔσονται ἑκατέρα ἑκατέρᾳ, ὑφ᾽
ἃς αἱ ἴσαι πλευραὶ ὑποτείνουσιν.

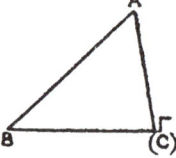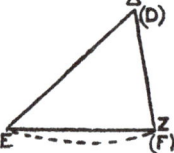

ΕΣΤΩ δύο τρίγωνα τὰ ΑΒΓ, ΔΕΖ τὰς δύο πλευρὰς
10   τὰς ΑΒ, ΑΓ ταῖς δυσὶ πλευραῖς ταῖς ΔΕ, ΔΖ
ἴσας ἔχοντα ἑκατέραν ἑκατέρᾳ τὴν μὲν ΑΒ τῇ ΔΕ
τὴν δὲ ΑΓ τῇ ΔΖ καὶ γωνίαν τὴν ὑπὸ ΒΑΓ γωνίᾳ
τῇ ὑπὸ ΕΔΖ ἴσην. λέγω ὅτι καὶ βάσις ἡ ΒΓ βάσει
τῇ ΕΖ ἴση ἐστίν, καὶ τὸ ΑΒΓ τρίγωνον τῷ ΔΕΖ
15   τριγώνῳ ἴσον ἔσται, καὶ αἱ λοιπαὶ γωνίαι ταῖς λοιπαῖς γωνίαις ἴσαι ἔσονται ἑκατέρα ἑκατέρᾳ, ὑφ᾽ ἃς
αἱ ἴσαι πλευραὶ ὑποτείνουσιν, ἡ μὲν ὑπὸ ΑΒΓ τῇ
ὑπὸ ΔΕΖ, ἡ δὲ ὑπὸ ΑΓΒ τῇ ὑπὸ ΔΖΕ.
Ἐφαρμοζομένου γὰρ τοῦ ΑΒΓ τριγώνου ἐπὶ τὸ
20   ΔΕΖ τρίγωνον καὶ τιθεμένου τοῦ μὲν Α σημείου ἐπὶ

τὸ Δ σημεῖον τῆς δὲ ΑΒ εὐθείας ἐπὶ τὴν ΔΕ, ἐφαρμόσει καὶ τὸ Β σημεῖον ἐπὶ τὸ Ε διὰ τὸ ἴσην εἶναι τὴν ΑΒ τῇ ΔΕ· ἐφαρμοσάσης δὴ τῆς ΑΒ ἐπὶ τὴν ΔΕ ἐφαρμόσει καὶ ἡ ΑΓ εὐθεῖα ἐπὶ τὴν ΔΖ διὰ τὸ ἴσην εἶναι τὴν ὑπὸ ΒΑΓ γωνίαν τῇ ὑπὸ ΕΔΖ· ὥστε 25 καὶ τὸ Γ σημεῖον ἐπὶ τὸ Ζ σημεῖον ἐφαρμόσει διὰ τὸ ἴσην πάλιν εἶναι τὴν ΑΓ τῇ ΔΖ. ἀλλὰ μὴν καὶ τὸ Β ἐπὶ τὸ Ε ἐφηρμόκει· ὥστε βάσις ἡ ΒΓ ἐπὶ βάσιν τὴν ΕΖ ἐφαρμόσει. εἰ γὰρ τοῦ μὲν Β ἐπὶ τὸ Ε ἐφαρμόσαντος τοῦ δὲ Γ ἐπὶ τὸ Ζ ἡ ΒΓ βάσις ἐπὶ 30 τὴν ΕΖ οὐκ ἐφαρμόσει, δύο εὐθεῖαι χωρίον περιέξουσιν· ὅπερ ἐστὶν ἀδύνατον. ἐφαρμόσει ἄρα ἡ ΒΓ βάσις ἐπὶ τὴν ΕΖ καὶ ἴση αὐτῇ ἔσται· ὥστε καὶ ὅλον τὸ ΑΒΓ τρίγωνον ἐπὶ ὅλον τὸ ΔΕΖ τρίγωνον ἐφαρμόσει καὶ ἴσον αὐτῷ ἔσται, καὶ αἱ λοιπαὶ γωνίαι 35 ἐπὶ τὰς λοιπὰς γωνίας ἐφαρμόσουσι καὶ ἴσαι αὐταῖς ἔσονται, ἡ μὲν ὑπὸ ΑΒΓ τῇ ὑπὸ ΔΕΖ ἡ δὲ ὑπὸ ΑΓΒ τῇ ὑπὸ ΔΖΕ.

Ἐὰν ἄρα δύο τρίγωνα τὰς δύο πλευρὰς [ταῖς] δύο πλευραῖς ἴσας ἔχῃ ἑκατέραν ἑκατέρᾳ καὶ τὴν γωνίαν 40 τῇ γωνίᾳ ἴσην ἔχῃ τὴν ὑπὸ τῶν ἴσων εὐθειῶν περιεχομένην, καὶ τὴν βάσιν τῇ βάσει ἴσην ἕξει, καὶ τὸ τρίγωνον τῷ τριγώνῳ ἴσον ἔσται, καὶ αἱ λοιπαὶ γωνίαι ταῖς λοιπαῖς γωνίαις ἴσαι ἔσονται ἑκατέρα ἑκατέρᾳ, ὑφ᾽ ἃς αἱ ἴσαι πλευραὶ ὑποτείνουσιν· ὅπερ 45 ἔδει δεῖξαι.

ε'.

Τῶν ἰσοσκελῶν τριγώνων αἱ πρὸς τῇ βάσει γωνίαι ἴσαι ἀλλήλαις εἰσίν, καὶ προσεκβληθεισῶν τῶν ἴσων εὐθειῶν αἱ ὑπὸ τὴν βάσιν γωνίαι ἴσαι ἀλλήλαις ἔσονται.

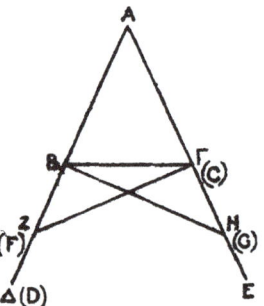

5 ΕΣΤΩ τρίγωνον ἰσοσκελὲς τὸ ΑΒΓ ἴσην ἔχον τὴν ΑΒ πλευρὰν τῇ ΑΓ πλευρᾷ, καὶ προσεκβεβλήσθωσαν ἐπ' εὐθείας ταῖς ΑΒ, ΑΓ εὐθεῖαι αἱ ΒΔ, ΓΕ· λέγω ὅτι ἡ μὲν ὑπὸ ΑΒΓ γωνία τῇ ὑπὸ ΑΓΒ ἴση ἐστίν, ἡ δὲ ὑπὸ ΓΒΔ τῇ ὑπὸ ΒΓΕ.

10 εἰλήφθω γὰρ ἐπὶ τῆς ΒΔ τυχὸν σημεῖον τὸ Ζ, καὶ ἀφῃρήσθω ἀπὸ τῆς μείζονος τῆς ΑΕ τῇ ἐλάσσονι τῇ ΑΖ ἴση ἡ ΑΗ, καὶ ἐπεζεύχθωσαν αἱ ΖΓ, ΗΒ εὐθεῖαι.

ἐπεὶ οὖν ἴση ἐστὶν ἡ μὲν ΑΖ τῇ ΑΗ ἡ δὲ ΑΒ τῇ ΑΓ, δύο δὴ αἱ ΖΑ, ΑΓ δυσὶ ταῖς ΗΑ, ΑΒ ἴσαι 15 εἰσὶν ἑκατέρα ἑκατέρᾳ· καὶ γωνίαν κοινὴν περιέχουσι τὴν ὑπὸ ΖΑΗ· βάσις ἄρα ἡ ΖΓ βάσει τῇ ΗΒ ἴση

ἐστίν, καὶ τὸ ΑΖΓ τρίγωνον τῷ ΑΗΒ τριγώνῳ ἴσον
ἔσται, καὶ αἱ λοιπαὶ γωνίαι ταῖς λοιπαῖς γωνίαις ἴσαι
ἔσονται ἑκατέρα ἑκατέρᾳ, ὑφ' ἃς αἱ ἴσαι πλευραὶ
ὑποτείνουσιν, ἡ μὲν ὑπὸ ΑΓΖ τῇ ὑπὸ ΑΒΗ, ἡ δὲ 20
ὑπὸ ΑΖΓ τῇ ὑπὸ ΑΗΒ. καὶ ἐπεὶ ὅλη ἡ ΑΖ ὅλῃ τῇ
ΑΗ ἐστιν ἴση, ὧν ἡ ΑΒ τῇ ΑΓ ἐστιν ἴση, λοιπὴ
ἄρα ἡ ΒΖ λοιπῇ τῇ ΓΗ ἐστιν ἴση. ἐδείχθη δὲ καὶ
ἡ ΖΓ τῇ ΗΒ ἴση· δύο δὴ αἱ ΒΖ, ΖΓ δυσὶ ταῖς ΓΗ,
ΗΒ ἴσαι εἰσὶν ἑκατέρα ἑκατέρᾳ· καὶ γωνία ἡ ὑπὸ 25
ΒΖΓ γωνίᾳ τῇ ὑπὸ ΓΗΒ ἴση, καὶ βάσις αὐτῶν
κοινὴ ἡ ΒΓ· καὶ τὸ ΒΖΓ ἄρα τρίγωνον τῷ ΓΗΒ
τριγώνῳ ἴσον ἔσται, καὶ αἱ λοιπαὶ γωνίαι ταῖς λοιπαῖς γωνίαις ἴσαι ἔσονται ἑκατέρα ἑκατέρᾳ, ὑφ' ἃς
αἱ ἴσαι πλευραὶ ὑποτείνουσιν· ἴση ἄρα ἐστὶν ἡ μὲν 30
ὑπὸ ΖΒΓ τῇ ὑπὸ ΗΓΒ ἡ δὲ ὑπὸ ΒΓΖ τῇ ὑπὸ ΓΒΗ.
ἐπεὶ οὖν ὅλη ἡ ὑπὸ ΑΒΗ γωνία ὅλῃ τῇ ὑπὸ ΑΓΖ
γωνίᾳ ἐδείχθη ἴση, ὧν ἡ ὑπὸ ΓΒΗ τῇ ὑπὸ ΒΓΖ ἴση,
λοιπὴ ἄρα ἡ ὑπὸ ΑΒΓ λοιπῇ τῇ ὑπὸ ΑΓΒ ἐστιν
ἴση· καί εἰσι πρὸς τῇ βάσει τοῦ ΑΒΓ τριγώνου. 35
ἐδείχθη δὲ καὶ ἡ ὑπὸ ΖΒΓ τῇ ὑπὸ ΗΓΒ ἴση· καί
εἰσιν ὑπὸ τὴν βάσιν.

Τῶν ἄρα ἰσοσκελῶν τριγώνων αἱ πρὸς τῇ βάσει
γωνίαι ἴσαι ἀλλήλαις εἰσίν, καὶ προσεκβληθεισῶν
τῶν ἴσων εὐθειῶν αἱ ὑπὸ τὴν βάσιν γωνίαι ἴσαι 40
ἀλλήλαις ἔσονται· ὅπερ ἔδει δεῖξαι.

ϛ΄.

Ἐὰν τριγώνου αἱ δύο γωνίαι ἴσαι ἀλλήλαις
ὦσιν, καὶ αἱ ὑπὸ τὰς ἴσας γωνίας ὑποτείνουσαι
πλευραὶ ἴσαι ἀλλήλαις ἔσονται.

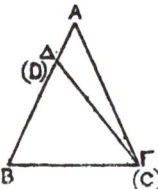

ΕΣΤΩ τρίγωνον τὸ ΑΒΓ ἴσην ἔχον τὴν ὑπὸ ΑΒΓ
5   γωνίαν τῇ ὑπὸ ΑΓΒ γωνίᾳ· λέγω ὅτι καὶ πλευρὰ
ἡ ΑΒ πλευρᾷ τῇ ΑΓ ἐστιν ἴση.

εἰ γὰρ ἄνισός ἐστιν ἡ ΑΒ τῇ ΑΓ, ἡ ἑτέρα αὐτῶν
μείζων ἐστίν. ἔστω μείζων ἡ ΑΒ, καὶ ἀφῃρήσθω
ἀπὸ τῆς μείζονος τῆς ΑΒ τῇ ἐλάττονι τῇ ΑΓ ἴση ἡ
10   ΔΒ, καὶ ἐπεζεύχθω ἡ ΔΓ.

Ἐπεὶ οὖν ἴση ἐστὶν ἡ ΔΒ τῇ ΑΓ κοινὴ δὲ ἡ ΒΓ,
δύο δὴ αἱ ΔΒ, ΒΓ δύο ταῖς ΑΓ, ΓΒ ἴσαι εἰσὶν
ἑκατέρα ἑκατέρᾳ, καὶ γωνία ἡ ὑπὸ ΔΒΓ γωνίᾳ τῇ
ὑπὸ ΑΓΒ ἐστιν ἴση· βάσις ἄρα ἡ ΔΓ βάσει τῇ ΑΒ
15   ἴση ἐστίν, καὶ τὸ ΔΒΓ τρίγωνον τῷ ΑΓΒ τριγώνῳ
ἴσον ἔσται, τὸ ἔλασσον τῷ μείζονι· ὅπερ ἄτοπον·
οὐκ ἄρα ἄνισός ἐστιν ἡ ΑΒ τῇ ΑΓ· ἴση ἄρα.

Ἐὰν ἄρα τριγώνου αἱ δύο γωνίαι ἴσαι ἀλλήλαις
ὦσιν, καὶ αἱ ὑπὸ τὰς ἴσας γωνίας ὑποτείνουσαι
20   πλευραὶ ἴσαι ἀλλήλαις ἔσονται· ὅπερ ἔδει δεῖξαι.

ζ'.

Ἐπὶ τῆς αὐτῆς εὐθείας δύο ταῖς αὐταῖς εὐθείαις
ἄλλαι δύο εὐθεῖαι ἴσαι ἑκατέρα ἑκατέρᾳ οὐ
συσταθήσονται πρὸς ἄλλῳ καὶ ἄλλῳ σημείῳ
ἐπὶ τὰ αὐτὰ μέρη τὰ αὐτὰ πέρατα ἔχουσαι
ταῖς ἐξ ἀρχῆς εὐθείαις.　5

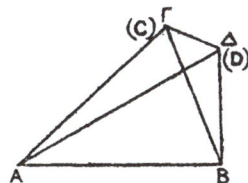

Εἰ γὰρ δυνατόν, ἐπὶ τῆς αὐτῆς εὐθείας τῆς ΑΒ
δύο ταῖς αὐταῖς εὐθείαις ταῖς ΑΓ, ΓΒ ἄλλαι δύο
εὐθεῖαι αἱ ΑΔ, ΔΒ ἴσαι ἑκατέρα ἑκατέρᾳ συνεστάτωσαν πρὸς ἄλλῳ καὶ ἄλλῳ σημείῳ τῷ τε Γ καὶ Δ
ἐπὶ τὰ αὐτὰ μέρη τὰ αὐτὰ πέρατα ἔχουσαι, ὥστε　10
ἴσην εἶναι τὴν μὲν ΓΑ τῇ ΔΑ τὸ αὐτὸ πέρας ἔχουσαν
αὐτῇ τὸ Α, τὴν δὲ ΓΒ τῇ ΔΒ τὸ αὐτὸ πέρας ἔχουσαν
αὐτῇ τὸ Β, καὶ ἐπεζεύχθω ἡ ΓΔ.

Ἐπεὶ οὖν ἴση ἐστὶν ἡ ΑΓ τῇ ΑΔ, ἴση ἐστὶ καὶ
γωνία ἡ ὑπὸ ΑΓΔ τῇ ὑπὸ ΑΔΓ· μείζων ἄρα ἡ ὑπὸ　15
ΑΔΓ τῆς ὑπὸ ΔΓΒ· πολλῷ ἄρα ἡ ὑπὸ ΓΔΒ μείζων
ἐστὶ τῆς ὑπὸ ΔΓΒ. πάλιν ἐπεὶ ἴση ἐστὶν ἡ ΓΒ τῇ
ΔΒ, ἴση ἐστὶ καὶ γωνία ἡ ὑπὸ ΓΔΒ γωνίᾳ τῇ ὑπὸ
ΔΓΒ. ἐδείχθη δὲ αὐτῆς καὶ πολλῷ μείζων· ὅπερ
ἐστὶν ἀδύνατον.　20

Οὐκ ἄρα ἐπὶ τῆς αὐτῆς εὐθείας δύο ταῖς αὐταῖς

εὐθείαις ἄλλαι δύο εὐθεῖαι ἴσαι ἑκατέρα ἑκατέρᾳ
συσταθήσονται πρὸς ἄλλῳ καὶ ἄλλῳ σημείῳ ἐπὶ τὰ
αὐτὰ μέρη τὰ αὐτὰ πέρατα ἔχουσαι ταῖς ἐξ ἀρχῆς
25 εὐθείαις· ὅπερ ἔδει δεῖξαι.

η΄.

Ἐὰν δύο τρίγωνα τὰς δύο πλευρὰς [ταῖς] δύο
πλευραῖς ἴσας ἔχῃ ἑκατέραν ἑκατέρᾳ, ἔχῃ δὲ
καὶ τὴν βάσιν τῇ βάσει ἴσην, καὶ τὴν γωνίαν
τῇ γωνίᾳ ἴσην ἕξει τὴν ὑπὸ τῶν ἴσων εὐθειῶν
5 περιεχομένην.

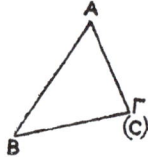 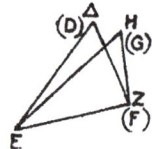

ΕΣΤΩ δύο τρίγωνα τὰ ΑΒΓ, ΔΕΖ τὰς δύο πλευρὰς
τὰς ΑΒ, ΑΓ ταῖς δύο πλευραῖς ταῖς ΔΕ, ΔΖ
ἴσας ἔχοντα ἑκατέραν ἑκατέρᾳ, τὴν μὲν ΑΒ τῇ ΔΕ
τὴν δὲ ΑΓ τῇ ΔΖ· ἐχέτω δὲ καὶ βάσιν τὴν ΒΓ
10 βάσει τῇ ΕΖ ἴσην· λέγω ὅτι καὶ γωνία ἡ ὑπὸ ΒΑΓ
γωνίᾳ τῇ ὑπὸ ΕΔΖ ἐστιν ἴση.

Ἐφαρμοζομένου γὰρ τοῦ ΑΒΓ τριγώνου ἐπὶ τὸ
ΔΕΖ τρίγωνον καὶ τιθεμένου τοῦ μὲν Β σημείου ἐπὶ
τὸ Ε σημεῖον τῆς δὲ ΒΓ εὐθείας ἐπὶ τὴν ΕΖ ἐφαρ-
15 μόσει καὶ τὸ Γ σημεῖον ἐπὶ τὸ Ζ διὰ τὸ ἴσην εἶναι

τὴν ΒΓ τῇ ΕΖ· ἐφαρμοσάσης δὴ τῆς ΒΓ ἐπὶ τὴν
ΕΖ ἐφαρμόσουσι καὶ αἱ ΒΑ, ΓΑ ἐπὶ τὰς ΕΔ, ΔΖ.
εἰ γὰρ βάσις μὲν ἡ ΒΓ ἐπὶ βάσιν τὴν ΕΖ ἐφαρμόσει, αἱ δὲ ΒΑ, ΑΓ πλευραὶ ἐπὶ τὰς ΕΔ, ΔΖ οὐκ
ἐφαρμόσουσιν ἀλλὰ παραλλάξουσιν ὡς αἱ ΕΗ, ΗΖ, 20
συσταθήσονται ἐπὶ τῆς αὐτῆς εὐθείας δύο ταῖς αὐταῖς
εὐθείαις ἄλλαι δύο εὐθεῖαι ἴσαι ἑκατέρα ἑκατέρᾳ πρὸς
ἄλλῳ καὶ ἄλλῳ σημείῳ ἐπὶ τὰ αὐτὰ μέρη τὰ αὐτὰ
πέρατα ἔχουσαι. οὐ συνίστανται δέ· οὐκ ἄρα ἐφαρ-
μοζομένης τῆς ΒΓ βάσεως ἐπὶ τὴν ΕΖ βάσιν οὐκ 25
ἐφαρμόσουσι καὶ αἱ ΒΑ, ΑΓ πλευραὶ ἐπὶ τὰς ΕΔ,
ΔΖ. ἐφαρμόσουσιν ἄρα· ὥστε καὶ γωνία ἡ ὑπὸ ΒΑΓ
ἐπὶ γωνίαν τὴν ὑπὸ ΕΔΖ ἐφαρμόσει καὶ ἴση αὐτῇ
ἔσται.

Ἐὰν ἄρα δύο τρίγωνα τὰς δύο πλευρὰς [ταῖς] δύο 30
πλευραῖς ἴσας ἔχῃ ἑκατέραν ἑκατέρᾳ καὶ τὴν βάσιν
τῇ βάσει ἴσην ἔχῃ, καὶ τὴν γωνίαν τῇ γωνίᾳ ἴσην
ἕξει τὴν ὑπὸ τῶν ἴσων εὐθειῶν περιεχομένην· ὅπερ
ἔδει δεῖξαι.

θ΄.

Τὴν δοθεῖσαν γωνίαν εὐθύγραμμον δίχα τεμεῖν.

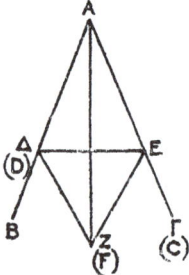

ΕΣΤΩ ἡ δοθεῖσα γωνία εὐθύγραμμος ἡ ὑπὸ ΒΑΓ. δεῖ δὴ αὐτὴν δίχα τεμεῖν.

Εἰλήφθω ἐπὶ τῆς ΑΒ τυχὸν σημεῖον τὸ Δ, καὶ
5 ἀφῃρήσθω ἀπὸ τῆς ΑΓ τῇ ΑΔ ἴση ἡ ΑΕ, καὶ ἐπεζεύχθω ἡ ΔΕ, καὶ συνεστάτω ἐπὶ τῆς ΔΕ τρίγωνον
ἰσόπλευρον τὸ ΔΕΖ, καὶ ἐπεζεύχθω ἡ ΑΖ· λέγω
ὅτι ἡ ὑπὸ ΒΑΓ γωνία δίχα τέτμηται ὑπὸ τῆς ΑΖ
εὐθείας.

10 Ἐπεὶ γὰρ ἴση ἐστὶν ἡ ΑΔ τῇ ΑΕ, κοινὴ δὲ ἡ
ΑΖ, δύο δὴ αἱ ΔΑ, ΑΖ δυσὶ ταῖς ΕΑ, ΑΖ ἴσαι εἰσὶν
ἑκατέρα ἑκατέρᾳ. καὶ βάσις ἡ ΔΖ βάσει τῇ ΕΖ ἴση
ἐστίν· γωνία ἄρα ἡ ὑπὸ ΔΑΖ γωνίᾳ τῇ ὑπὸ ΕΑΖ
ἴση ἐστίν.

15 Ἡ ἄρα δοθεῖσα γωνία εὐθύγραμμος ἡ ὑπὸ ΒΑΓ
δίχα τέτμηται ὑπὸ τῆς ΑΖ εὐθείας· ὅπερ ἔδει ποιῆσαι.

ι΄.

Τὴν δοθεῖσαν εὐθεῖαν πεπερασμένην δίχα τεμεῖν.

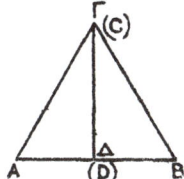

Ε ΣΤΩ ἡ δοθεῖσα εὐθεῖα πεπερασμένη ἡ ΑΒ· δεῖ
δὴ τὴν ΑΒ εὐθεῖαν πεπερασμένην δίχα τεμεῖν.

Συνεστάτω ἐπ᾽ αὐτῆς τρίγωνον ἰσόπλευρον τὸ
ΑΒΓ, καὶ τετμήσθω ἡ ὑπὸ ΑΓΒ γωνία δίχα τῇ ΓΔ 5
εὐθείᾳ· λέγω ὅτι ἡ ΑΒ εὐθεῖα δίχα τέτμηται κατὰ
τὸ Δ σημεῖον.

Ἐπεὶ γὰρ ἴση ἐστὶν ἡ ΑΓ τῇ ΓΒ, κοινὴ δὲ ἡ
ΓΔ, δύο δὴ αἱ ΑΓ, ΓΔ δύο ταῖς ΒΓ, ΓΔ ἴσαι εἰσὶν
ἑκατέρα ἑκατέρᾳ· καὶ γωνία ἡ ὑπὸ ΑΓΔ γωνίᾳ τῇ 10
ὑπὸ ΒΓΔ ἴση ἐστίν· βάσις ἄρα ἡ ΑΔ βάσει τῇ ΒΔ
ἴση ἐστίν.

Ἡ ἄρα δοθεῖσα εὐθεῖα πεπερασμένη ἡ ΑΒ δίχα
τέτμηται κατὰ τὸ Δ· ὅπερ ἔδει ποιῆσαι.

ια΄.

Τῇ δοθείσῃ εὐθείᾳ ἀπὸ τοῦ πρὸς αὐτῇ δοθέντος
σημείου πρὸς ὀρθὰς γωνίας εὐθεῖαν γραμμὴν
ἀγαγεῖν.

Ε ΣΤΩ ἡ μὲν δοθεῖσα εὐθεῖα ἡ ΑΒ τὸ δὲ δοθὲν
σημεῖον ἐπ᾽ αὐτῆς τὸ Γ· δεῖ δὴ ἀπὸ τοῦ Γ 5

σημείου τῇ ΑΒ εὐθείᾳ πρὸς ὀρθὰς γωνίας εὐθεῖαν γραμμὴν ἀγαγεῖν.

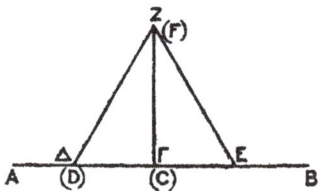

Εἰλήφθω ἐπὶ τῆς ΑΓ τυχὸν σημεῖον τὸ Δ, καὶ κείσθω τῇ ΓΔ ἴση ἡ ΓΕ, καὶ συνεστάτω ἐπὶ τῆς ΔΕ
10 τρίγωνον ἰσόπλευρον τὸ ΖΔΕ, καὶ ἐπεζεύχθω ἡ ΖΓ· λέγω ὅτι τῇ δοθείσῃ εὐθείᾳ τῇ ΑΒ ἀπὸ τοῦ πρὸς αὐτῇ δοθέντος σημείου τοῦ Γ πρὸς ὀρθὰς γωνίας εὐθεῖα γραμμὴ ἦκται ἡ ΖΓ.

Ἐπεὶ γὰρ ἴση ἐστὶν ἡ ΔΓ τῇ ΓΕ, κοινὴ δὲ ἡ
15 ΓΖ, δύο δὴ αἱ ΔΓ, ΓΖ δυσὶ ταῖς ΕΓ, ΓΖ ἴσαι εἰσὶν ἑκατέρα ἑκατέρᾳ· καὶ βάσις ἡ ΔΖ βάσει τῇ ΖΕ ἴση ἐστίν· γωνία ἄρα ἡ ὑπὸ ΔΓΖ γωνίᾳ τῇ ὑπὸ ΕΓΖ ἴση ἐστίν· καί εἰσιν ἐφεξῆς. ὅταν δὲ εὐθεῖα ἐπ᾽ εὐθεῖαν σταθεῖσα τὰς ἐφεξῆς γωνίας ἴσας ἀλλήλαις ποιῇ,
20 ὀρθὴ ἑκατέρα τῶν ἴσων γωνιῶν ἐστιν· ὀρθὴ ἄρα ἐστὶν ἑκατέρα τῶν ὑπὸ ΔΓΖ, ΖΓΕ.

Τῇ ἄρα δοθείσῃ εὐθείᾳ τῇ ΑΒ ἀπὸ τοῦ πρὸς αὐτῇ δοθέντος σημείου τοῦ Γ πρὸς ὀρθὰς γωνίας εὐθεῖα γραμμὴ ἦκται ἡ ΓΖ· ὅπερ ἔδει ποιῆσαι.

ιβ'.

Ἐπὶ τὴν δοθεῖσαν εὐθεῖαν ἄπειρον ἀπὸ τοῦ
δοθέντος σημείου, ὃ μή ἐστιν ἐπ' αὐτῆς,
κάθετον εὐθεῖαν γραμμὴν ἀγαγεῖν.

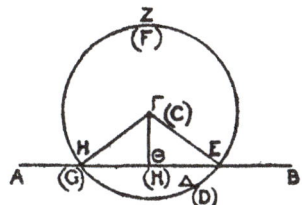

ΕΣΤΩ ἡ μὲν δοθεῖσα εὐθεῖα ἄπειρος ἡ ΑΒ τὸ δὲ
δοθὲν σημεῖον, ὃ μή ἐστιν ἐπ' αὐτῆς, τὸ Γ· δεῖ 5
δὴ ἐπὶ τὴν δοθεῖσαν εὐθεῖαν ἄπειρον τὴν ΑΒ ἀπὸ
τοῦ δοθέντος σημείου τοῦ Γ, ὃ μή ἐστιν ἐπ' αὐτῆς,
κάθετον εὐθεῖαν γραμμὴν ἀγαγεῖν.

Εἰλήφθω γὰρ ἐπὶ τὰ ἕτερα μέρη τῆς ΑΒ εὐθείας
τυχὸν σημεῖον τὸ Δ, καὶ κέντρῳ μὲν τῷ Γ διαστήματι 10
δὲ τῷ ΓΔ κύκλος γεγράφθω ὁ ΕΖΗ, καὶ τετμήσθω
ἡ ΕΗ εὐθεῖα δίχα κατὰ τὸ Θ, καὶ ἐπεζεύχθωσαν αἱ
ΓΗ, ΓΘ, ΓΕ εὐθεῖαι· λέγω ὅτι ἐπὶ τὴν δοθεῖσαν
εὐθεῖαν ἄπειρον τὴν ΑΒ ἀπὸ τοῦ δοθέντος σημείου
τοῦ Γ, ὃ μή ἐστιν ἐπ' αὐτῆς, κάθετος ἦκται ἡ ΓΘ. 15

Ἐπεὶ γὰρ ἴση ἐστὶν ἡ ΗΘ τῇ ΘΕ, κοινὴ δὲ ἡ
ΘΓ, δύο δὴ αἱ ΗΘ, ΘΓ δύο ταῖς ΕΘ, ΘΓ ἴσαι εἰσὶν
ἑκατέρα ἑκατέρᾳ· καὶ βάσις ἡ ΓΗ βάσει τῇ ΓΕ ἐστιν
ἴση· γωνία ἄρα ἡ ὑπὸ ΓΘΗ γωνίᾳ τῇ ὑπὸ ΕΘΓ ἐστιν

20 ἴση. καί εἰσιν ἐφεξῆς. ὅταν δὲ εὐθεῖα ἐπ᾽ εὐθεῖαν
σταθεῖσα τὰς ἐφεξῆς γωνίας ἴσας ἀλλήλαις ποιῇ,
ὀρθὴ ἑκατέρα τῶν ἴσων γωνιῶν ἐστίν, καὶ ἡ ἐφεστηκυῖα εὐθεῖα κάθετος καλεῖται ἐφ᾽ ἣν ἐφέστηκεν.
Ἐπὶ τὴν δοθεῖσαν ἄρα εὐθεῖαν ἄπειρον τὴν ΑΒ
25 ἀπὸ τοῦ δοθέντος σημείου τοῦ Γ, ὃ μή ἐστιν ἐπ᾽
αὐτῆς, κάθετος ἦκται ἡ ΓΘ· ὅπερ ἔδει ποιῆσαι.

ιγ'.

Ἐὰν εὐθεῖα ἐπ᾽ εὐθεῖαν σταθεῖσα γωνίας ποιῇ,
ἤτοι δύο ὀρθὰς ἢ δυσὶν ὀρθαῖς ἴσας ποιήσει.

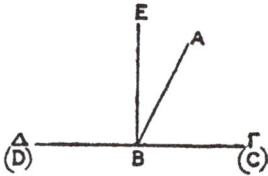

ΕΥΘΕΙΑ γάρ τις ἡ ΑΒ ἐπ᾽ εὐθεῖαν τὴν ΓΔ σταθεῖσα γωνίας ποιείτω τὰς ὑπὸ ΓΒΑ, ΑΒΔ· λέγω
5 ὅτι αἱ ὑπὸ ΓΒΑ, ΑΒΔ γωνίαι ἤτοι δύο ὀρθαί εἰσιν ἢ
δυσὶν ὀρθαῖς ἴσαι.

Εἰ μὲν οὖν ἴση ἐστὶν ἡ ὑπὸ ΓΒΑ τῇ ὑπὸ ΑΒΔ,
δύο ὀρθαί εἰσιν. εἰ δὲ οὔ, ἤχθω ἀπὸ τοῦ Β σημείου
τῇ ΓΔ [εὐθείᾳ] πρὸς ὀρθὰς ἡ ΒΕ· αἱ ἄρα ὑπὸ ΓΒΕ,
10 ΕΒΔ δύο ὀρθαί εἰσιν· καὶ ἐπεὶ ἡ ὑπὸ ΓΒΕ δυσὶ ταῖς
ὑπὸ ΓΒΑ, ΑΒΕ ἴση ἐστίν, κοινὴ προσκείσθω ἡ ὑπὸ
ΕΒΔ· αἱ ἄρα ὑπὸ ΓΒΕ, ΕΒΔ τρισὶ ταῖς ὑπὸ ΓΒΑ,

ΑΒΕ, ΕΒΔ ἴσαι εἰσίν. πάλιν, ἐπεὶ ἡ ὑπὸ ΔΒΑ δυσὶ
ταῖς ὑπὸ ΔΒΕ, ΕΒΑ ἴση ἐστίν, κοινὴ προσκείσθω ἡ
ὑπὸ ΑΒΓ· αἱ ἄρα ὑπὸ ΔΒΑ, ΑΒΓ τρισὶ ταῖς ὑπὸ 15
ΔΒΕ, ΕΒΑ, ΑΒΓ ἴσαι εἰσίν. ἐδείχθησαν δὲ καὶ αἱ
ὑπὸ ΓΒΕ, ΕΒΔ τρισὶ ταῖς αὐταῖς ἴσαι· τὰ δὲ τῷ
αὐτῷ ἴσα καὶ ἀλλήλοις ἐστὶν ἴσα· καὶ αἱ ὑπὸ ΓΒΕ,
ΕΒΔ ἄρα ταῖς ὑπὸ ΔΒΑ, ΑΒΓ ἴσαι εἰσίν· ἀλλὰ αἱ
ὑπὸ ΓΒΕ, ΕΒΔ δύο ὀρθαί εἰσιν· καὶ αἱ ὑπὸ ΔΒΑ, 20
ΑΒΓ ἄρα δυσὶν ὀρθαῖς ἴσαι εἰσίν.

Ἐὰν ἄρα εὐθεῖα ἐπ' εὐθεῖαν σταθεῖσα γωνίας ποιῇ,
ἤτοι δύο ὀρθὰς ἢ δυσὶν ὀρθαῖς ἴσας ποιήσει· ὅπερ
ἔδει δεῖξαι.

ιδ'.

Ἐὰν πρός τινι εὐθείᾳ καὶ τῷ πρὸς αὐτῇ σημείῳ
δύο εὐθεῖαι μὴ ἐπὶ τὰ αὐτὰ μέρη κείμεναι
τὰς ἐφεξῆς γωνίας δυσὶν ὀρθαῖς ἴσας ποιῶσιν,
ἐπ' εὐθείας ἔσονται ἀλλήλαις αἱ εὐθεῖαι.

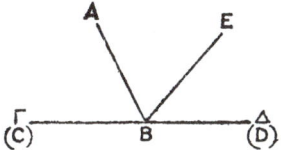

ΠΡΟΣ γάρ τινι εὐθείᾳ τῇ ΑΒ καὶ τῷ πρὸς αὐτῇ 5
σημείῳ τῷ Β δύο εὐθεῖαι αἱ ΒΓ, ΒΔ μὴ ἐπὶ τὰ
αὐτὰ μέρη κείμεναι τὰς ἐφεξῆς γωνίας τὰς ὑπὸ ΑΒΓ,
ΑΒΔ δύο ὀρθαῖς ἴσας ποιείτωσαν· λέγω ὅτι ἐπ'
εὐθείας ἐστὶ τῇ ΓΒ ἡ ΒΔ.

10 Εἰ γὰρ μή ἐστι τῇ ΒΓ ἐπ' εὐθείας ἡ ΒΔ, ἔστω τῇ ΓΒ ἐπ' εὐθείας ἡ ΒΕ.

Ἐπεὶ οὖν εὐθεῖα ἡ ΑΒ ἐπ' εὐθεῖαν τὴν ΓΒΕ ἐφέστηκεν, αἱ ἄρα ὑπὸ ΑΒΓ, ΑΒΕ γωνίαι δύο ὀρθαῖς ἴσαι εἰσίν· εἰσὶ δὲ καὶ αἱ ὑπὸ ΑΒΓ, ΑΒΔ δύο ὀρθαῖς 15 ἴσαι· αἱ ἄρα ὑπὸ ΓΒΑ, ΑΒΕ ταῖς ὑπὸ ΓΒΑ, ΑΒΔ ἴσαι εἰσίν. κοινὴ ἀφῃρήσθω ἡ ὑπὸ ΓΒΑ· λοιπὴ ἄρα ἡ ὑπὸ ΑΒΕ λοιπῇ τῇ ὑπὸ ΑΒΔ ἐστιν ἴση, ἡ ἐλάσσων τῇ μείζονι· ὅπερ ἐστὶν ἀδύνατον. οὐκ ἄρα ἐπ' εὐθείας ἐστὶν ἡ ΒΕ τῇ ΓΒ. ὁμοίως δὴ δείξομεν 20 ὅτι οὐδὲ ἄλλη τις πλὴν τῆς ΒΔ· ἐπ' εὐθείας ἄρα ἐστὶν ἡ ΓΒ τῇ ΒΔ.

Ἐὰν ἄρα πρός τινι εὐθείᾳ καὶ τῷ πρὸς αὐτῇ σημείῳ δύο εὐθεῖαι μὴ ἐπὶ τὰ αὐτὰ μέρη κείμεναι τὰς ἐφεξῆς γωνίας δυσὶν ὀρθαῖς ἴσας ποιῶσιν, ἐπ' εὐθείας ἔσονται 25 ἀλλήλαις αἱ εὐθεῖαι· ὅπερ ἔδει δεῖξαι.

ιε'.

Ἐὰν δύο εὐθεῖαι τέμνωσιν ἀλλήλας, τὰς κατὰ κορυφὴν γωνίας ἴσας ἀλλήλαις ποιοῦσιν.

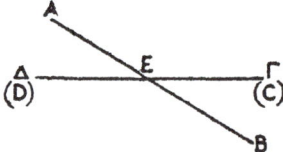

ΔΥΟ γὰρ εὐθεῖαι αἱ ΑΒ, ΓΔ τεμνέτωσαν ἀλλήλας κατὰ τὸ Ε σημεῖον· λέγω ὅτι ἴση ἐστὶν ἡ μὲν

ὑπὸ ΑΕΓ γωνία τῇ ὑπὸ ΔΕΒ, ἡ δὲ ὑπὸ ΓΕΒ τῇ 5
ὑπὸ ΑΕΔ.

Ἐπεὶ γὰρ εὐθεῖα ἡ ΑΕ ἐπ' εὐθεῖαν τὴν ΓΔ ἐφέστηκε
γωνίας ποιοῦσα τὰς ὑπὸ ΓΕΑ, ΑΕΔ, αἱ ἄρα ὑπὸ
ΓΕΑ, ΑΕΔ γωνίαι δυσὶν ὀρθαῖς ἴσαι εἰσίν. πάλιν,
ἐπεὶ εὐθεῖα ἡ ΔΕ ἐπ' εὐθεῖαν τὴν ΑΒ ἐφέστηκε 10
γωνίας ποιοῦσα τὰς ὑπὸ ΑΕΔ, ΔΕΒ, αἱ ἄρα ὑπὸ
ΑΕΔ, ΔΕΒ γωνίαι δυσὶν ὀρθαῖς ἴσαι εἰσίν. ἐδείχθησαν δὲ καὶ αἱ ὑπὸ ΓΕΑ, ΑΕΔ δυσὶν ὀρθαῖς ἴσαι·
αἱ ἄρα ὑπὸ ΓΕΑ, ΑΕΔ ταῖς ὑπὸ ΑΕΔ, ΔΕΒ ἴσαι
εἰσίν. κοινὴ ἀφῃρήσθω ἡ ὑπὸ ΑΕΔ· λοιπὴ ἄρα ἡ 15
ὑπὸ ΓΕΑ λοιπῇ τῇ ὑπὸ ΒΕΔ ἴση ἐστίν. ὁμοίως δὴ
δειχθήσεται ὅτι καὶ αἱ ὑπὸ ΓΕΒ, ΔΕΑ ἴσαι εἰσίν.

Ἐὰν ἄρα δύο εὐθεῖαι τέμνωσιν ἀλλήλας, τὰς κατὰ
κορυφὴν γωνίας ἴσας ἀλλήλαις ποιοῦσιν· ὅπερ ἔδει
δεῖξαι. 20

[Πόρισμα.

Ἐκ δὴ τούτου φανερὸν ὅτι, ἐὰν δύο εὐθεῖαι τέμνωσιν ἀλλήλας, τὰς πρὸς τῇ τομῇ γωνίας τέτρασιν
ὀρθαῖς ἴσας ποιήσουσιν.]

ιϛ'.

Παντὸς τριγώνου μιᾶς τῶν πλευρῶν προσεκ-
βληθείσης ἡ ἐκτὸς γωνία ἑκατέρας τῶν ἐντὸς
καὶ ἀπεναντίον γωνιῶν μείζων ἐστίν.

ΕΣΤΩ τρίγωνον τὸ ΑΒΓ, καὶ προσεκβεβλήσθω
αὐτοῦ μία πλευρὰ ἡ ΒΓ ἐπὶ τὸ Δ· λέγω ὅτι ἡ 5

ἐκτὸς γωνία ἡ ὑπὸ ΑΓΔ μείζων ἐστὶν ἑκατέρας τῶν
ἐντὸς καὶ ἀπεναντίον τῶν ὑπὸ ΓΒΑ, ΒΑΓ γωνιῶν.

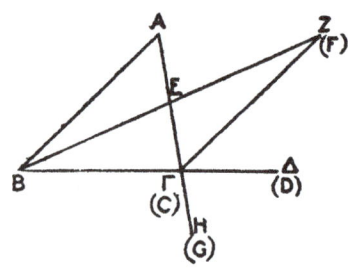

Τετμήσθω ἡ ΑΓ δίχα κατὰ τὸ Ε, καὶ ἐπιζευχθεῖσα
ἡ ΒΕ ἐκβεβλήσθω ἐπ᾽ εὐθείας ἐπὶ τὸ Ζ, καὶ κείσθω
10 τῇ ΒΕ ἴση ἡ ΕΖ, καὶ ἐπεζεύχθω ἡ ΖΓ, καὶ διήχθω
ἡ ΑΓ ἐπὶ τὸ Η.
Ἐπεὶ οὖν ἴση ἐστὶν ἡ μὲν ΑΕ τῇ ΕΓ, ἡ δὲ ΒΕ
τῇ ΕΖ, δύο δὴ αἱ ΑΕ, ΕΒ δυσὶ ταῖς ΓΕ, ΕΖ ἴσαι
εἰσὶν ἑκατέρα ἑκατέρᾳ· καὶ γωνία ἡ ὑπὸ ΑΕΒ γωνίᾳ
15 τῇ ὑπὸ ΖΕΓ ἴση ἐστίν· κατὰ κορυφὴν γάρ· βάσις
ἄρα ἡ ΑΒ βάσει τῇ ΖΓ ἴση ἐστίν, καὶ τὸ ΑΒΕ
τρίγωνον τῷ ΖΕΓ τριγώνῳ ἐστὶν ἴσον, καὶ αἱ λοιπαὶ
γωνίαι ταῖς λοιπαῖς γωνίαις ἴσαι εἰσὶν ἑκατέρα ἑκατέρᾳ, ὑφ᾽ ἃς αἱ ἴσαι πλευραὶ ὑποτείνουσιν· ἴση ἄρα
20 ἐστὶν ἡ ὑπὸ ΒΑΕ τῇ ὑπὸ ΕΓΖ. μείζων δέ ἐστιν ἡ
ὑπὸ ΕΓΔ τῆς ὑπὸ ΕΓΖ· μείζων ἄρα ἡ ὑπὸ ΑΓΔ
τῆς ὑπὸ ΒΑΕ. Ὁμοίως δὴ τῆς ΒΓ τετμημένης δίχα
δειχθήσεται καὶ ἡ ὑπὸ ΒΓΗ, τουτέστιν ἡ ὑπὸ ΑΓΔ,
μείζων καὶ τῆς ὑπὸ ΑΒΓ.
25   Παντὸς ἄρα τριγώνου μιᾶς τῶν πλευρῶν προσεκ-

βληθείσης ἡ ἐκτὸς γωνία ἑκατέρας τῶν ἐντὸς καὶ
ἀπεναντίον γωνιῶν μείζων ἐστίν· ὅπερ ἔδει δεῖξαι.

ιζ΄.

Παντὸς τριγώνου αἱ δύο γωνίαι δύο ὀρθῶν
ἐλάσσονές εἰσι πάντη μεταλαμβανόμεναι.

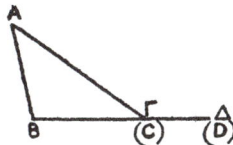

ΕΣΤΩ τρίγωνον τὸ ΑΒΓ· λέγω ὅτι τοῦ ΑΒΓ
τριγώνου αἱ δύο γωνίαι δύο ὀρθῶν ἐλάττονές
εἰσι πάντη μεταλαμβανόμεναι.   5
Ἐκβεβλήσθω γὰρ ἡ ΒΓ ἐπὶ τὸ Δ.
Καὶ ἐπεὶ τριγώνου τοῦ ΑΒΓ ἐκτός ἐστι γωνία ἡ
ὑπὸ ΑΓΔ, μείζων ἐστὶ τῆς ἐντὸς καὶ ἀπεναντίον τῆς
ὑπὸ ΑΒΓ. κοινὴ προσκείσθω ἡ ὑπὸ ΑΓΒ· αἱ ἄρα
ὑπὸ ΑΓΔ, ΑΓΒ τῶν ὑπὸ ΑΒΓ, ΒΓΑ μείζονές εἰσιν. 10
ἀλλ᾽ αἱ ὑπὸ ΑΓΔ, ΑΓΒ δύο ὀρθαῖς ἴσαι εἰσίν· αἱ
ἄρα ὑπὸ ΑΒΓ, ΒΓΑ δύο ὀρθῶν ἐλάσσονές εἰσιν.
ὁμοίως δὴ δείξομεν ὅτι καὶ αἱ ὑπὸ ΒΑΓ, ΑΓΒ δύο
ὀρθῶν ἐλάσσονές εἰσι καὶ ἔτι αἱ ὑπὸ ΓΑΒ, ΑΒΓ.
Παντὸς ἄρα τριγώνου αἱ δύο γωνίαι δύο ὀρθῶν 15
ἐλάσσονές εἰσι πάντη μεταλαμβανόμεναι· ὅπερ ἔδει
δεῖξαι.

ιη΄.

Παντὸς τριγώνου ἡ μείζων πλευρὰ τὴν μείζονα
γωνίαν ὑποτείνει.

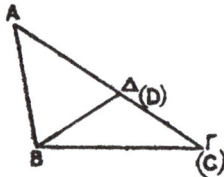

ΕΣΤΩ γὰρ τρίγωνον τὸ ΑΒΓ μείζονα ἔχον τὴν
ΑΓ πλευρὰν τῆς ΑΒ· λέγω ὅτι καὶ γωνία ἡ
5 ὑπὸ ΑΒΓ μείζων ἐστὶ τῆς ὑπὸ ΒΓΑ.

Ἐπεὶ γὰρ μείζων ἐστὶν ἡ ΑΓ τῆς ΑΒ, κείσθω τῇ
ΑΒ ἴση ἡ ΑΔ, καὶ ἐπεζεύχθω ἡ ΒΔ.

Καὶ ἐπεὶ τριγώνου τοῦ ΒΓΔ ἐκτός ἐστι γωνία ἡ
ὑπὸ ΑΔΒ, μείζων ἐστὶ τῆς ἐντὸς καὶ ἀπεναντίον τῆς
10 ὑπὸ ΔΓΒ· ἴση δὲ ἡ ὑπὸ ΑΔΒ τῇ ὑπὸ ΑΒΔ, ἐπεὶ
καὶ πλευρὰ ἡ ΑΒ τῇ ΑΔ ἐστιν ἴση· μείζων ἄρα καὶ
ἡ ὑπὸ ΑΒΔ τῆς ὑπὸ ΑΓΒ· πολλῷ ἄρα ἡ ὑπὸ ΑΒΓ
μείζων ἐστὶ τῆς ὑπὸ ΑΓΒ.

Παντὸς ἄρα τριγώνου ἡ μείζων πλευρὰ τὴν μείζονα
15 γωνίαν ὑποτείνει· ὅπερ ἔδει δεῖξαι.

ιθ'.

Παντὸς τριγώνου ὑπὸ τὴν μείζονα γωνίαν ἡ
μείζων πλευρὰ ὑποτείνει.

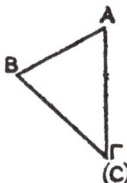

Ε ΣΤΩ τρίγωνον τὸ ΑΒΓ μείζονα ἔχον τὴν ὑπὸ
ΑΒΓ γωνίαν τῆς ὑπὸ ΒΓΑ· λέγω ὅτι καὶ πλευρὰ
ἡ ΑΓ πλευρᾶς τῆς ΑΒ μείζων ἐστίν.   5

Εἰ γὰρ μή, ἤτοι ἴση ἐστὶν ἡ ΑΓ τῇ ΑΒ ἢ ἐλάσ-
σων· ἴση μὲν οὖν οὐκ ἔστιν ἡ ΑΓ τῇ ΑΒ· ἴση γὰρ
ἂν ἦν καὶ γωνία ἡ ὑπὸ ΑΒΓ τῇ ὑπὸ ΑΓΒ· οὐκ ἔστι
δέ· οὐκ ἄρα ἴση ἐστὶν ἡ ΑΓ τῇ ΑΒ. οὐδὲ μὴν
ἐλάσσων ἐστὶν ἡ ΑΓ τῆς ΑΒ· ἐλάσσων γὰρ ἂν ἦν 10
καὶ γωνία ἡ ὑπὸ ΑΒΓ τῆς ὑπὸ ΑΓΒ· οὐκ ἔστι δέ·
οὐκ ἄρα ἐλάσσων ἐστὶν ἡ ΑΓ τῆς ΑΒ. ἐδείχθη δὲ
ὅτι οὐδὲ ἴση ἐστίν. μείζων ἄρα ἐστὶν ἡ ΑΓ τῆς ΑΒ.

Παντὸς ἄρα τριγώνου ὑπὸ τὴν μείζονα γωνίαν ἡ
μείζων πλευρὰ ὑποτείνει· ὅπερ ἔδει δεῖξαι.   15

κ'.

Παντὸς τριγώνου αἱ δύο πλευραὶ τῆς λοιπῆς
μείζονές εἰσι πάντῃ μεταλαμβανόμεναι.

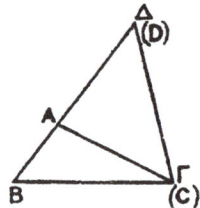

ΕΣΤΩ γὰρ τρίγωνον τὸ ΑΒΓ· λέγω ὅτι τοῦ ΑΒΓ
τριγώνου αἱ δύο πλευραὶ τῆς λοιπῆς μείζονές
5 εἰσι πάντῃ μεταλαμβανόμεναι, αἱ μὲν ΒΑ, ΑΓ τῆς
ΒΓ, αἱ δὲ ΑΒ, ΒΓ τῆς ΑΓ, αἱ δὲ ΒΓ, ΓΑ τῆς ΑΒ.
Διήχθω γὰρ ἡ ΒΑ ἐπὶ τὸ Δ σημεῖον, καὶ κείσθω
τῇ ΓΑ ἴση ἡ ΑΔ, καὶ ἐπεζεύχθω ἡ ΔΓ.
Ἐπεὶ οὖν ἴση ἐστὶν ἡ ΔΑ τῇ ΑΓ, ἴση ἐστὶ καὶ
10 γωνία ἡ ὑπὸ ΑΔΓ τῇ ὑπὸ ΑΓΔ· μείζων ἄρα ἡ ὑπὸ
ΒΓΔ τῆς ὑπὸ ΑΔΓ· καὶ ἐπεὶ τρίγωνόν ἐστι τὸ ΔΓΒ
μείζονα ἔχον τὴν ὑπὸ ΒΓΔ γωνίαν τῆς ὑπὸ ΒΔΓ,
ὑπὸ δὲ τὴν μείζονα γωνίαν ἡ μείζων πλευρὰ ὑποτείνει,
ἡ ΔΒ ἄρα τῆς ΒΓ ἐστι μείζων. ἴση δὲ ἡ ΔΑ τῇ
15 ΑΓ· μείζονες ἄρα αἱ ΒΑ, ΑΓ τῆς ΒΓ. ὁμοίως δὴ
δείξομεν ὅτι καὶ αἱ μὲν ΑΒ, ΒΓ τῆς ΓΑ μείζονές
εἰσιν, αἱ δὲ ΒΓ, ΓΑ τῆς ΑΒ.
Παντὸς ἄρα τριγώνου αἱ δύο πλευραὶ τῆς λοιπῆς
μείζονές εἰσι πάντῃ μεταλαμβανόμεναι· ὅπερ ἔδει
20 δεῖξαι.

κα'.

Ἐὰν τριγώνου ἐπὶ μιᾶς τῶν πλευρῶν ἀπὸ τῶν
περάτων δύο εὐθεῖαι ἐντὸς συσταθῶσιν, αἱ
συσταθεῖσαι τῶν λοιπῶν τοῦ τριγώνου δύο
πλευρῶν ἐλάττονες μὲν ἔσονται, μείζονα δὲ
γωνίαν περιέξουσιν.   5

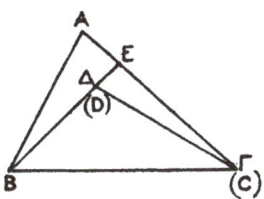

ΤΡΙΓΩΝΟΥ γὰρ τοῦ ΑΒΓ ἐπὶ μιᾶς τῶν πλευρῶν
τῆς ΒΓ ἀπὸ τῶν περάτων τῶν Β, Γ δύο εὐθεῖαι
ἐντὸς συνεστάτωσαν αἱ ΒΔ, ΔΓ· λέγω ὅτι αἱ ΒΔ,
ΔΓ τῶν λοιπῶν τοῦ τριγώνου δύο πλευρῶν τῶν ΒΑ,
ΑΓ ἐλάσσονες μέν εἰσιν, μείζονα δὲ γωνίαν περι-   10
έχουσι τὴν ὑπὸ ΒΔΓ τῆς ὑπὸ ΒΑΓ.
    Διήχθω γὰρ ἡ ΒΔ ἐπὶ τὸ Ε. καὶ ἐπεὶ παντὸς
τριγώνου αἱ δύο πλευραὶ τῆς λοιπῆς μείζονές εἰσιν,
τοῦ ΑΒΕ ἄρα τριγώνου αἱ δύο πλευραὶ αἱ ΑΒ, ΑΕ
τῆς ΒΕ μείζονές εἰσιν· κοινὴ προσκείσθω ἡ ΕΓ· αἱ 15
ἄρα ΒΑ, ΑΓ τῶν ΒΕ, ΕΓ μείζονές εἰσιν. πάλιν,
ἐπεὶ τοῦ ΓΕΔ τριγώνου αἱ δύο πλευραὶ αἱ ΓΕ, ΕΔ
τῆς ΓΔ μείζονές εἰσιν, κοινὴ προσκείσθω ἡ ΔΒ· αἱ
ΓΕ, ΕΒ ἄρα τῶν ΓΔ, ΔΒ μείζονές εἰσιν. ἀλλὰ
τῶν ΒΕ, ΕΓ μείζονες ἐδείχθησαν αἱ ΒΑ, ΑΓ· πολλῷ 20
ἄρα αἱ ΒΑ, ΑΓ τῶν ΒΔ, ΔΓ μείζονές εἰσιν.
    Πάλιν, ἐπεὶ παντὸς τριγώνου ἡ ἐκτὸς γωνία τῆς

ἐντὸς καὶ ἀπεναντίον μείζων ἐστίν, τοῦ ΓΔΕ ἄρα τριγώνου ἡ ἐκτὸς γωνία ἡ ὑπὸ ΒΔΓ μείζων ἐστὶ τῆς 25 ὑπὸ ΓΕΔ. διὰ ταῦτα τοίνυν καὶ τοῦ ΑΒΕ τριγώνου ἡ ἐκτὸς γωνία ἡ ὑπὸ ΓΕΒ μείζων ἐστὶ τῆς ὑπὸ ΒΑΓ. ἀλλὰ τῆς ὑπὸ ΓΕΒ μείζων ἐδείχθη ἡ ὑπὸ ΒΔΓ· πολλῷ ἄρα ἡ ὑπὸ ΒΔΓ μείζων ἐστὶ τῆς ὑπὸ ΒΑΓ. Ἐὰν ἄρα τριγώνου ἐπὶ μιᾶς τῶν πλευρῶν ἀπὸ τῶν 30 περάτων δύο εὐθεῖαι ἐντὸς συσταθῶσιν, αἱ συσταθεῖσαι τῶν λοιπῶν τοῦ τριγώνου δύο πλευρῶν ἐλάττονες μέν εἰσιν, μείζονα δὲ γωνίαν περιέχουσιν· ὅπερ ἔδει δεῖξαι.

κβ'.

Ἐκ τριῶν εὐθειῶν, αἵ εἰσιν ἴσαι τρισὶ ταῖς δοθείσαις [εὐθείαις], τρίγωνον συστήσασθαι· δεῖ δὴ τὰς δύο τῆς λοιπῆς μείζονας εἶναι πάντῃ μεταλαμβανομένας [διὰ τὸ καὶ παντὸς 5 τριγώνου τὰς δύο πλευρὰς τῆς λοιπῆς μείζονας εἶναι πάντῃ μεταλαμβανομένας].

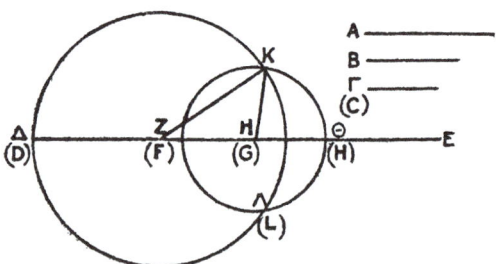

ΕΣΤΩΣΑΝ αἱ δοθεῖσαι τρεῖς εὐθεῖαι αἱ Α, Β, Γ, ὧν αἱ δύο τῆς λοιπῆς μείζονες ἔστωσαν πάντῃ

μεταλαμβανόμεναι, αἱ μὲν Α, Β τῆς Γ, αἱ δὲ Α, Γ
τῆς Β, καὶ ἔτι αἱ Β, Γ τῆς Α· δεῖ δὴ ἐκ τῶν ἴσων 10
ταῖς Α, Β, Γ τρίγωνον συστήσασθαι.

Ἐκκείσθω τις εὐθεῖα ἡ ΔΕ πεπερασμένη μὲν κατὰ
τὸ Δ ἄπειρος δὲ κατὰ τὸ Ε, καὶ κείσθω τῇ μὲν Α
ἴση ἡ ΔΖ, τῇ δὲ Β ἴση ἡ ΖΗ, τῇ δὲ Γ ἴση ἡ ΗΘ·
καὶ κέντρῳ μὲν τῷ Ζ, διαστήματι δὲ τῷ ΖΔ κύκλος 15
γεγράφθω ὁ ΔΚΛ· πάλιν κέντρῳ μὲν τῷ Η, διαστήματι δὲ τῷ ΗΘ κύκλος γεγράφθω ὁ ΚΛΘ, καὶ ἐπεζεύχθωσαν αἱ ΚΖ, ΚΗ· λέγω ὅτι ἐκ τριῶν εὐθειῶν
τῶν ἴσων ταῖς Α, Β, Γ τρίγωνον συνέσταται τὸ ΚΖΗ.

Ἐπεὶ γὰρ τὸ Ζ σημεῖον κέντρον ἐστὶ τοῦ ΔΚΛ 20
κύκλου, ἴση ἐστὶν ἡ ΖΔ τῇ ΖΚ· ἀλλὰ ἡ ΖΔ τῇ Α
ἐστιν ἴση. καὶ ἡ ΚΖ ἄρα τῇ Α ἐστιν ἴση. πάλιν,
ἐπεὶ τὸ Η σημεῖον κέντρον ἐστὶ τοῦ ΛΚΘ κύκλου,
ἴση ἐστὶν ἡ ΗΘ τῇ ΗΚ· ἀλλὰ ἡ ΗΘ τῇ Γ ἐστιν
ἴση· καὶ ἡ ΚΗ ἄρα τῇ Γ ἐστιν ἴση. ἐστὶ δὲ καὶ ἡ 25
ΖΗ τῇ Β ἴση· αἱ τρεῖς ἄρα εὐθεῖαι αἱ ΚΖ, ΖΗ, ΗΚ
τρισὶ ταῖς Α, Β, Γ ἴσαι εἰσίν.

Ἐκ τριῶν ἄρα εὐθειῶν τῶν ΚΖ, ΖΗ, ΗΚ, αἵ εἰσιν
ἴσαι τρισὶ ταῖς δοθείσαις εὐθείαις ταῖς Α, Β, Γ, τρίγωνον συνέσταται τὸ ΚΖΗ· ὅπερ ἔδει ποιῆσαι. 30

κγ'.

Πρὸς τῇ δοθείσῃ εὐθείᾳ καὶ τῷ πρὸς αὐτῇ σημείῳ
τῇ δοθείσῃ γωνίᾳ εὐθυγράμμῳ ἴσην γωνίαν
εὐθύγραμμον συστήσασθαι.

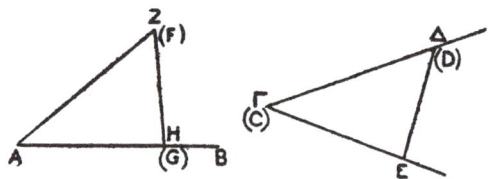

5 ΕΣΤΩ ἡ μὲν δοθεῖσα εὐθεῖα ἡ ΑΒ, τὸ δὲ πρὸς
αὐτῇ σημεῖον τὸ Α, ἡ δὲ δοθεῖσα γωνία εὐθύγραμμος ἡ ὑπὸ ΔΓΕ· δεῖ δὴ πρὸς τῇ δοθείσῃ εὐθείᾳ
τῇ ΑΒ καὶ τῷ πρὸς αὐτῇ σημείῳ τῷ Α τῇ δοθείσῃ
γωνίᾳ εὐθυγράμμῳ τῇ ὑπὸ ΔΓΕ ἴσην γωνίαν εὐθύγραμμον συστήσασθαι.

10 Εἰλήφθω ἐφ' ἑκατέρας τῶν ΓΔ, ΓΕ τυχόντα σημεῖα τὰ Δ, Ε, καὶ ἐπεζεύχθω ἡ ΔΕ· καὶ ἐκ τριῶν
εὐθειῶν, αἵ εἰσιν ἴσαι τρισὶ ταῖς ΓΔ, ΔΕ, ΓΕ, τρίγωνον συνεστάτω τὸ ΑΖΗ, ὥστε ἴσην εἶναι τὴν μὲν
ΓΔ τῇ ΑΖ, τὴν δὲ ΓΕ τῇ ΑΗ, καὶ ἔτι τὴν ΔΕ τῇ ΖΗ.

15 Ἐπεὶ οὖν δύο αἱ ΔΓ, ΓΕ δύο ταῖς ΖΑ, ΑΗ ἴσαι
εἰσὶν ἑκατέρα ἑκατέρᾳ, καὶ βάσις ἡ ΔΕ βάσει τῇ
ΖΗ ἴση, γωνία ἄρα ἡ ὑπὸ ΔΓΕ γωνίᾳ τῇ ὑπὸ ΖΑΗ
ἐστιν ἴση.

Πρὸς ἄρα τῇ δοθείσῃ εὐθείᾳ τῇ ΑΒ καὶ τῷ πρὸς
20 αὐτῇ σημείῳ τῷ Α τῇ δοθείσῃ γωνίᾳ εὐθυγράμμῳ
τῇ ὑπὸ ΔΓΕ ἴση γωνία εὐθύγραμμος συνέσταται ἡ
ὑπὸ ΖΑΗ· ὅπερ ἔδει ποιῆσαι.

κδ'.

Ἐὰν δύο τρίγωνα τὰς δύο πλευρὰς [ταῖς] δύο
πλευραῖς ἴσας ἔχῃ ἑκατέραν ἑκατέρᾳ, τὴν δὲ
γωνίαν τῆς γωνίας μείζονα ἔχῃ τὴν ὑπὸ τῶν
ἴσων εὐθειῶν περιεχομένην, καὶ τὴν βάσιν
τῆς βάσεως μείζονα ἕξει.   5

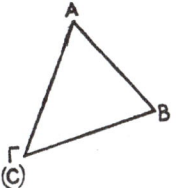 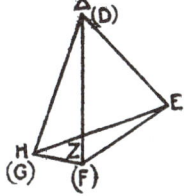

ΕΣΤΩ δύο τρίγωνα τὰ ΑΒΓ, ΔΕΖ τὰς δύο πλευρὰς
τὰς ΑΒ, ΑΓ ταῖς δύο πλευραῖς ταῖς ΔΕ, ΔΖ
ἴσας ἔχοντα ἑκατέραν ἑκατέρᾳ, τὴν μὲν ΑΒ τῇ ΔΕ
τὴν δὲ ΑΓ τῇ ΔΖ, ἡ δὲ πρὸς τῷ Α γωνία τῆς πρὸς
τῷ Δ γωνίας μείζων ἔστω· λέγω ὅτι καὶ βάσις ἡ 10
ΒΓ βάσεως τῆς ΕΖ μείζων ἐστίν.

Ἐπεὶ γὰρ μείζων ἡ ὑπὸ ΒΑΓ γωνία τῆς ὑπὸ ΕΔΖ
γωνίας, συνεστάτω πρὸς τῇ ΔΕ εὐθείᾳ καὶ τῷ πρὸς
αὐτῇ σημείῳ τῷ Δ τῇ ὑπὸ ΒΑΓ γωνίᾳ ἴση ἡ ὑπὸ
ΕΔΗ, καὶ κείσθω ὁποτέρᾳ τῶν ΑΓ, ΔΖ ἴση ἡ ΔΗ, 15
καὶ ἐπεζεύχθωσαν αἱ ΕΗ, ΖΗ.

Ἐπεὶ οὖν ἴση ἐστὶν ἡ μὲν ΑΒ τῇ ΔΕ, ἡ δὲ ΑΓ
τῇ ΔΗ, δύο δὴ αἱ ΒΑ, ΑΓ δυσὶ ταῖς ΕΔ, ΔΗ ἴσαι
εἰσὶν ἑκατέρα ἑκατέρᾳ· καὶ γωνία ἡ ὑπὸ ΒΑΓ γωνίᾳ
τῇ ὑπὸ ΕΔΗ ἴση· βάσις ἄρα ἡ ΒΓ βάσει τῇ ΕΗ 20
ἐστιν ἴση. πάλιν, ἐπεὶ ἴση ἐστὶν ἡ ΔΖ τῇ ΔΗ, ἴση

ἐστὶ καὶ ἡ ὑπὸ ΔΗΖ γωνία τῇ ὑπὸ ΔΖΗ· μείζων
ἄρα ἡ ὑπὸ ΔΖΗ τῆς ὑπὸ ΕΗΖ· πολλῷ ἄρα μείζων
ἐστὶν ἡ ὑπὸ ΕΖΗ τῆς ὑπὸ ΕΗΖ. καὶ ἐπεὶ τρίγωνόν
25 ἐστι τὸ ΕΖΗ μείζονα ἔχον τὴν ὑπὸ ΕΖΗ γωνίαν
τῆς ὑπὸ ΕΗΖ, ὑπὸ δὲ τὴν μείζονα γωνίαν ἡ μείζων
πλευρὰ ὑποτείνει, μείζων ἄρα καὶ πλευρὰ ἡ ΕΗ τῆς
ΕΖ. ἴση δὲ ἡ ΕΗ τῇ ΒΓ· μείζων ἄρα καὶ ἡ ΒΓ
τῆς ΕΖ.
30 Ἐὰν ἄρα δύο τρίγωνα τὰς δύο πλευρὰς δυσὶ
πλευραῖς ἴσας ἔχῃ ἑκατέραν ἑκατέρᾳ, τὴν δὲ γωνίαν
τῆς γωνίας μείζονα ἔχῃ τὴν ὑπὸ τῶν ἴσων εὐθειῶν
περιεχομένην, καὶ τὴν βάσιν τῆς βάσεως μείζονα
ἕξει· ὅπερ ἔδει δεῖξαι.

κε΄.

Ἐὰν δύο τρίγωνα τὰς δύο πλευρὰς δυσὶ πλευραῖς
ἴσας ἔχῃ ἑκατέραν ἑκατέρᾳ, τὴν δὲ βάσιν τῆς
βάσεως μείζονα ἔχῃ, καὶ τὴν γωνίαν τῆς
γωνίας μείζονα ἕξει τὴν ὑπὸ τῶν ἴσων εὐθειῶν
5 περιεχομένην.

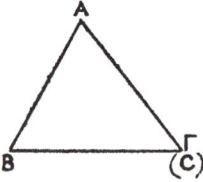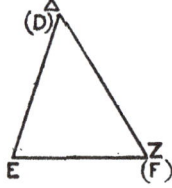

ΕΣΤΩ δύο τρίγωνα τὰ ΑΒΓ, ΔΕΖ τὰς δύο πλευρὰς
τὰς ΑΒ, ΑΓ ταῖς δύο πλευραῖς ταῖς ΔΕ, ΔΖ
ἴσας ἔχοντα ἑκατέραν ἑκατέρᾳ, τὴν μὲν ΑΒ τῇ ΔΕ,

τὴν δὲ ΑΓ τῇ ΔΖ· βάσις δὲ ἡ ΒΓ βάσεως τῆς ΕΖ
μείζων ἔστω· λέγω ὅτι καὶ γωνία ἡ ὑπὸ ΒΑΓ γωνίας 10
τῆς ὑπὸ ΕΔΖ μείζων ἐστίν.

Εἰ γὰρ μή, ἤτοι ἴση ἐστὶν αὐτῇ ἢ ἐλάσσων· ἴση
μὲν οὖν οὐκ ἔστιν ἡ ὑπὸ ΒΑΓ τῇ ὑπὸ ΕΔΖ· ἴση
γὰρ ἂν ἦν καὶ βάσις ἡ ΒΓ βάσει τῇ ΕΖ· οὐκ ἔστι
δέ. οὐκ ἄρα ἴση ἐστὶ γωνία ἡ ὑπὸ ΒΑΓ τῇ ὑπὸ 15
ΕΔΖ· οὐδὲ μὴν ἐλάσσων ἐστὶν ἡ ὑπὸ ΒΑΓ τῆς ὑπὸ
ΕΔΖ· ἐλάσσων γὰρ ἂν ἦν καὶ βάσις ἡ ΒΓ βάσεως
τῆς ΕΖ· οὐκ ἔστι δέ· οὐκ ἄρα ἐλάσσων ἐστὶν ἡ ὑπὸ
ΒΑΓ γωνία τῆς ὑπὸ ΕΔΖ. ἐδείχθη δὲ ὅτι οὐδὲ ἴση·
μείζων ἄρα ἐστὶν ἡ ὑπὸ ΒΑΓ τῆς ὑπὸ ΕΔΖ. 20

Ἐὰν ἄρα δύο τρίγωνα τὰς δύο πλευρὰς δυσὶ
πλευραῖς ἴσας ἔχῃ ἑκατέραν ἑκατέρᾳ, τὴν δὲ βάσιν
τῆς βάσεως μείζονα ἔχῃ, καὶ τὴν γωνίαν τῆς γωνίας
μείζονα ἕξει τὴν ὑπὸ τῶν ἴσων εὐθειῶν περιεχομένην·
ὅπερ ἔδει δεῖξαι. 25

κϛʹ.

Ἐὰν δύο τρίγωνα τὰς δύο γωνίας δυσὶ γωνίαις
ἴσας ἔχῃ ἑκατέραν ἑκατέρᾳ καὶ μίαν πλευρὰν
μιᾷ πλευρᾷ ἴσην ἤτοι τὴν πρὸς ταῖς ἴσαις
γωνίαις ἢ τὴν ὑποτείνουσαν ὑπὸ μίαν τῶν
ἴσων γωνιῶν, καὶ τὰς λοιπὰς πλευρὰς ταῖς 5
λοιπαῖς πλευραῖς ἴσας ἕξει [ἑκατέραν ἑκατέρᾳ]
καὶ τὴν λοιπὴν γωνίαν τῇ λοιπῇ γωνίᾳ.

Ἔστω δύο τρίγωνα τὰ ΑΒΓ, ΔΕΖ τὰς δύο γωνίας
τὰς ὑπὸ ΑΒΓ, ΒΓΑ δυσὶ ταῖς ὑπὸ ΔΕΖ, ΕΖΔ
ἴσας ἔχοντα ἑκατέραν ἑκατέρᾳ, τὴν μὲν ὑπὸ ΑΒΓ τῇ 10
ὑπὸ ΔΕΖ, τὴν δὲ ὑπὸ ΒΓΑ τῇ ὑπὸ ΕΖΔ· ἐχέτω δὲ

καὶ μίαν πλευρὰν μιᾷ πλευρᾷ ἴσην, πρότερον τὴν
πρὸς ταῖς ἴσαις γωνίαις τὴν ΒΓ τῇ ΕΖ· λέγω ὅτι

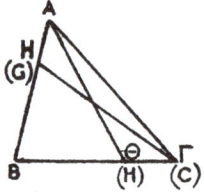 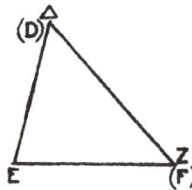

καὶ τὰς λοιπὰς πλευρὰς ταῖς λοιπαῖς πλευραῖς ἴσας
15 ἕξει ἑκατέραν ἑκατέρᾳ, τὴν μὲν ΑΒ τῇ ΔΕ τὴν δὲ
ΑΓ τῇ ΔΖ, καὶ τὴν λοιπὴν γωνίαν τῇ λοιπῇ γωνίᾳ,
τὴν ὑπὸ ΒΑΓ τῇ ὑπὸ ΕΔΖ.
    Εἰ γὰρ ἄνισός ἐστιν ἡ ΑΒ τῇ ΔΕ, μία αὐτῶν
μείζων ἐστίν. ἔστω μείζων ἡ ΑΒ, καὶ κείσθω τῇ
20 ΔΕ ἴση ἡ ΒΗ, καὶ ἐπεζεύχθω ἡ ΗΓ.
    Ἐπεὶ οὖν ἴση ἐστὶν ἡ μὲν ΒΗ τῇ ΔΕ, ἡ δὲ ΒΓ
τῇ ΕΖ, δύο δὴ αἱ ΒΗ, ΒΓ δυσὶ ταῖς ΔΕ, ΕΖ ἴσαι
εἰσὶν ἑκατέρα ἑκατέρᾳ· καὶ γωνία ἡ ὑπὸ ΗΒΓ γωνίᾳ
τῇ ὑπὸ ΔΕΖ ἴση ἐστίν· βάσις ἄρα ἡ ΗΓ βάσει τῇ
25 ΔΖ ἴση ἐστίν, καὶ τὸ ΗΒΓ τρίγωνον τῷ ΔΕΖ τριγώνῳ ἴσον ἐστίν, καὶ αἱ λοιπαὶ γωνίαι ταῖς λοιπαῖς
γωνίαις ἴσαι ἔσονται, ὑφ᾽ ἃς αἱ ἴσαι πλευραὶ ὑπο-
τείνουσιν· ἴση ἄρα ἡ ὑπὸ ΗΓΒ γωνία τῇ ὑπὸ ΔΖΕ.
ἀλλὰ ἡ ὑπὸ ΔΖΕ τῇ ὑπὸ ΒΓΑ ὑπόκειται ἴση· καὶ ἡ
30 ὑπὸ ΒΓΗ ἄρα τῇ ὑπὸ ΒΓΑ ἴση ἐστίν, ἡ ἐλάσσων
τῇ μείζονι· ὅπερ ἀδύνατον. οὐκ ἄρα ἄνισός ἐστιν ἡ
ΑΒ τῇ ΔΕ. ἴση ἄρα. ἔστι δὲ καὶ ἡ ΒΓ τῇ ΕΖ ἴση·
δύο δὴ αἱ ΑΒ, ΒΓ δυσὶ ταῖς ΔΕ, ΕΖ ἴσαι εἰσὶν
ἑκατέρα ἑκατέρᾳ· καὶ γωνία ἡ ὑπὸ ΑΒΓ γωνίᾳ τῇ

ὑπὸ ΔΕΖ ἐστιν ἴση· βάσις ἄρα ἡ ΑΓ βάσει τῇ ΔΖ 35
ἴση ἐστίν, καὶ λοιπὴ γωνία ἡ ὑπὸ ΒΑΓ τῇ λοιπῇ
γωνίᾳ τῇ ὑπὸ ΕΔΖ ἴση ἐστίν.

Ἀλλὰ δὴ πάλιν ἔστωσαν αἱ ὑπὸ τὰς ἴσας γωνίας
πλευραὶ ὑποτείνουσαι ἴσαι, ὡς ἡ ΑΒ τῇ ΔΕ· λέγω
πάλιν ὅτι καὶ αἱ λοιπαὶ πλευραὶ ταῖς λοιπαῖς πλευ- 40
ραῖς ἴσαι ἔσονται, ἡ μὲν ΑΓ τῇ ΔΖ, ἡ δὲ ΒΓ τῇ ΕΖ
καὶ ἔτι ἡ λοιπὴ γωνία ἡ ὑπὸ ΒΑΓ τῇ λοιπῇ γωνίᾳ
τῇ ὑπὸ ΕΔΖ ἴση ἐστίν.

Εἰ γὰρ ἄνισός ἐστιν ἡ ΒΓ τῇ ΕΖ, μία αὐτῶν
μείζων ἐστίν. ἔστω μείζων, εἰ δυνατόν, ἡ ΒΓ, καὶ 45
κείσθω τῇ ΕΖ ἴση ἡ ΒΘ, καὶ ἐπεζεύχθω ἡ ΑΘ. καὶ
ἐπεὶ ἴση ἐστὶν ἡ μὲν ΒΘ τῇ ΕΖ, ἡ δὲ ΑΒ τῇ ΔΕ,
δύο δὴ αἱ ΑΒ, ΒΘ δυσὶ ταῖς ΔΕ, ΕΖ ἴσαι εἰσὶν
ἑκατέρα ἑκατέρᾳ· καὶ γωνίας ἴσας περιέχουσιν· βάσις
ἄρα ἡ ΑΘ βάσει τῇ ΔΖ ἴση ἐστίν, καὶ τὸ ΑΒΘ 50
τρίγωνον τῷ ΔΕΖ τριγώνῳ ἴσον ἐστίν, καὶ αἱ λοιπαὶ
γωνίαι ταῖς λοιπαῖς γωνίαις ἴσαι ἔσονται, ὑφ' ἃς αἱ
ἴσαι πλευραὶ ὑποτείνουσιν· ἴση ἄρα ἐστὶν ἡ ὑπὸ
ΒΘΑ γωνία τῇ ὑπὸ ΕΖΔ. ἀλλὰ ἡ ὑπὸ ΕΖΔ τῇ
ὑπὸ ΒΓΑ ἐστιν ἴση· τριγώνου δὴ τοῦ ΑΘΓ ἡ ἐκτὸς 55
γωνία ἡ ὑπὸ ΒΘΑ ἴση ἐστὶ τῇ ἐντὸς καὶ ἀπεναντίον
τῇ ὑπὸ ΒΓΑ· ὅπερ ἀδύνατον. οὐκ ἄρα ἄνισός ἐστιν
ἡ ΒΓ τῇ ΕΖ· ἴση ἄρα. ἐστὶ δὲ καὶ ἡ ΑΒ τῇ ΔΕ
ἴση. δύο δὴ αἱ ΑΒ, ΒΓ δύο ταῖς ΔΕ, ΕΖ ἴσαι εἰσὶν
ἑκατέρα ἑκατέρᾳ· καὶ γωνίας ἴσας περιέχουσι· βάσις 60
ἄρα ἡ ΑΓ βάσει τῇ ΔΖ ἴση ἐστίν, καὶ τὸ ΑΒΓ
τρίγωνον τῷ ΔΕΖ τριγώνῳ ἴσον καὶ λοιπὴ γωνία ἡ
ὑπὸ ΒΑΓ τῇ λοιπῇ γωνίᾳ τῇ ὑπὸ ΕΔΖ ἴση.

Ἐὰν ἄρα δύο τρίγωνα τὰς δύο γωνίας δυσὶ γωνίαις

65 ἴσας ἔχῃ ἑκατέραν ἑκατέρᾳ καὶ μίαν πλευρὰν μιᾷ
πλευρᾷ ἴσην ἤτοι τὴν πρὸς ταῖς ἴσαις γωνίαις, ἢ τὴν
ὑποτείνουσαν ὑπὸ μίαν τῶν ἴσων γωνιῶν, καὶ τὰς
λοιπὰς πλευρὰς ταῖς λοιπαῖς πλευραῖς ἴσας ἕξει καὶ
τὴν λοιπὴν γωνίαν τῇ λοιπῇ γωνίᾳ· ὅπερ ἔδει δεῖξαι.

κζ'.

Ἐὰν εἰς δύο εὐθείας εὐθεῖα ἐμπίπτουσα τὰς ἐναλλὰξ γωνίας ἴσας ἀλλήλαις ποιῇ, παράλληλοι ἔσονται ἀλλήλαις αἱ εὐθεῖαι.

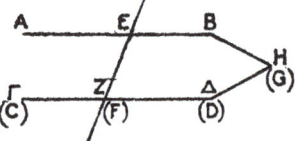

ΕΙΣ γὰρ δύο εὐθείας τὰς ΑΒ, ΓΔ εὐθεῖα ἐμπί-
5 πτουσα ἡ ΕΖ τὰς ἐναλλὰξ γωνίας τὰς ὑπὸ ΑΕΖ,
ΕΖΔ ἴσας ἀλλήλαις ποιείτω· λέγω ὅτι παράλληλός
ἐστιν ἡ ΑΒ τῇ ΓΔ.
Εἰ γὰρ μή, ἐκβαλλόμεναι αἱ ΑΒ, ΓΔ συμπεσοῦνται ἤτοι ἐπὶ τὰ Β, Δ μέρη ἢ ἐπὶ τὰ Α, Γ. ἐκβεβλή-
10 σθωσαν καὶ συμπιπτέτωσαν ἐπὶ τὰ Β, Δ μέρη κατὰ
τὸ Η. τριγώνου δὴ τοῦ ΗΕΖ ἡ ἐκτὸς γωνία ἡ ὑπὸ
ΑΕΖ ἴση ἐστὶ τῇ ἐντὸς καὶ ἀπεναντίον τῇ ὑπὸ ΕΖΗ·
ὅπερ ἐστὶν ἀδύνατον· οὐκ ἄρα αἱ ΑΒ, ΓΔ ἐκβαλλόμεναι συμπεσοῦνται ἐπὶ τὰ Β, Δ μέρη. ὁμοίως
15 δὴ δειχθήσεται ὅτι οὐδὲ ἐπὶ τὰ Α, Γ· αἱ δὲ ἐπὶ
μηδέτερα τὰ μέρη συμπίπτουσαι παράλληλοί εἰσιν·
παράλληλος ἄρα ἐστὶν ἡ ΑΒ τῇ ΓΔ.

Ἐὰν ἄρα εἰς δύο εὐθείας εὐθεῖα ἐμπίπτουσα τὰς
ἐναλλὰξ γωνίας ἴσας ἀλλήλαις ποιῇ, παράλληλοι
ἔσονται αἱ εὐθεῖαι· ὅπερ ἔδει δεῖξαι.   20

κη΄.

Ἐὰν εἰς δύο εὐθείας εὐθεῖα ἐμπίπτουσα τὴν
ἐκτὸς γωνίαν τῇ ἐντὸς καὶ ἀπεναντίον καὶ
ἐπὶ τὰ αὐτὰ μέρη ἴσην ποιῇ ἢ τὰς ἐντὸς καὶ
ἐπὶ τὰ αὐτὰ μέρη δυσὶν ὀρθαῖς ἴσας, παράλληλοι ἔσονται ἀλλήλαις αἱ εὐθεῖαι.   5

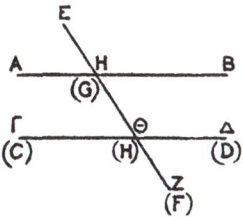

ΕΙΣ γὰρ δύο εὐθείας τὰς ΑΒ, ΓΔ εὐθεῖα ἐμπίπτουσα ἡ ΕΖ τὴν ἐκτὸς γωνίαν τὴν ὑπὸ ΕΗΒ
τῇ ἐντὸς καὶ ἀπεναντίον γωνίᾳ τῇ ὑπὸ ΗΘΔ ἴσην
ποιείτω ἢ τὰς ἐντὸς καὶ ἐπὶ τὰ αὐτὰ μέρη τὰς ὑπὸ
ΒΗΘ, ΗΘΔ δυσὶν ὀρθαῖς ἴσας· λέγω ὅτι παράλ-   10
ληλός ἐστιν ἡ ΑΒ τῇ ΓΔ.

Ἐπεὶ γὰρ ἴση ἐστὶν ἡ ὑπὸ ΕΗΒ τῇ ὑπὸ ΗΘΔ,
ἀλλὰ ἡ ὑπὸ ΕΗΒ τῇ ὑπὸ ΑΗΘ ἐστιν ἴση, καὶ ἡ
ὑπὸ ΑΗΘ ἄρα τῇ ὑπὸ ΗΘΔ ἐστιν ἴση· καί εἰσιν
ἐναλλάξ· παράλληλος ἄρα ἐστὶν ἡ ΑΒ τῇ ΓΔ.   15

Πάλιν, ἐπεὶ αἱ ὑπὸ ΒΗΘ, ΗΘΔ δύο ὀρθαῖς ἴσαι
εἰσίν, εἰσὶ δὲ καὶ αἱ ὑπὸ ΑΗΘ, ΒΗΘ δυσὶν ὀρθαῖς

H. E.   6

ἴσαι, αἱ ἄρα ὑπὸ ΑΗΘ, ΒΗΘ ταῖς ὑπὸ ΒΗΘ, ΗΘΔ
ἴσαι εἰσίν· κοινὴ ἀφῃρήσθω ἡ ὑπὸ ΒΗΘ· λοιπὴ ἄρα
20 ἡ ὑπὸ ΑΗΘ λοιπῇ τῇ ὑπὸ ΗΘΔ ἐστιν ἴση· καί εἰσιν
ἐναλλάξ· παράλληλος ἄρα ἐστὶν ἡ ΑΒ τῇ ΓΔ.
    Ἐὰν ἄρα εἰς δύο εὐθείας εὐθεῖα ἐμπίπτουσα τὴν
ἐκτὸς γωνίαν τῇ ἐντὸς καὶ ἀπεναντίον καὶ ἐπὶ τὰ
αὐτὰ μέρη ἴσην ποιῇ ἢ τὰς ἐντὸς καὶ ἐπὶ τὰ αὐτὰ
25 μέρη δυσὶν ὀρθαῖς ἴσας, παράλληλοι ἔσονται αἱ
εὐθεῖαι· ὅπερ ἔδει δεῖξαι.

κθ΄.

Ἡ εἰς τὰς παραλλήλους εὐθείας εὐθεῖα ἐμπίπτουσα τάς τε ἐναλλὰξ γωνίας ἴσας ἀλλήλαις
ποιεῖ καὶ τὴν ἐκτὸς τῇ ἐντὸς καὶ ἀπεναντίον
ἴσην καὶ τὰς ἐντὸς καὶ ἐπὶ τὰ αὐτὰ μέρη
5 δυσὶν ὀρθαῖς ἴσας.

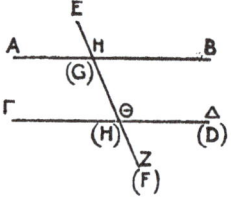

ΕΙΣ γὰρ παραλλήλους εὐθείας τὰς ΑΒ, ΓΔ εὐθεῖα
ἐμπιπτέτω ἡ ΕΖ· λέγω ὅτι τὰς ἐναλλὰξ γωνίας
τὰς ὑπὸ ΑΗΘ, ΗΘΔ ἴσας ποιεῖ καὶ τὴν ἐκτὸς γωνίαν
τὴν ὑπὸ ΕΗΒ τῇ ἐντὸς καὶ ἀπεναντίον τῇ ὑπὸ ΗΘΔ
10 ἴσην καὶ τὰς ἐντὸς καὶ ἐπὶ τὰ αὐτὰ μέρη τὰς ὑπὸ
ΒΗΘ, ΗΘΔ δυσὶν ὀρθαῖς ἴσας.

Εἰ γὰρ ἄνισός ἐστιν ἡ ὑπὸ ΑΗΘ τῇ ὑπὸ ΗΘΔ, μία αὐτῶν μείζων ἐστίν. ἔστω μείζων ἡ ὑπὸ ΑΗΘ· κοινὴ προσκείσθω ἡ ὑπὸ ΒΗΘ· αἱ ἄρα ὑπὸ ΑΗΘ, ΒΗΘ τῶν ὑπὸ ΒΗΘ, ΗΘΔ μείζονές εἰσιν. ἀλλὰ αἱ 15 ὑπὸ ΑΗΘ, ΒΗΘ δυσὶν ὀρθαῖς ἴσαι εἰσίν. [καὶ] αἱ ἄρα ὑπὸ ΒΗΘ, ΗΘΔ δύο ὀρθῶν ἐλάσσονές εἰσιν. αἱ δὲ ἀπ' ἐλασσόνων ἢ δύο ὀρθῶν ἐκβαλλόμεναι εἰς ἄπειρον συμπίπτουσιν· αἱ ἄρα ΑΒ, ΓΔ ἐκβαλλό-μεναι εἰς ἄπειρον συμπεσοῦνται· οὐ συμπίπτουσι δὲ 20 διὰ τὸ παραλλήλους αὐτὰς ὑποκεῖσθαι· οὐκ ἄρα ἄνισός ἐστιν ἡ ὑπὸ ΑΗΘ τῇ ὑπὸ ΗΘΔ· ἴση ἄρα. ἀλλὰ ἡ ὑπὸ ΑΗΘ τῇ ὑπὸ ΕΗΒ ἐστιν ἴση· καὶ ἡ ὑπὸ ΕΗΒ ἄρα τῇ ὑπὸ ΗΘΔ ἐστιν ἴση. κοινὴ προσκείσθω ἡ ὑπὸ ΒΗΘ· αἱ ἄρα ὑπὸ ΕΗΒ, ΒΗΘ 25 ταῖς ὑπὸ ΒΗΘ, ΗΘΔ ἴσαι εἰσίν. ἀλλὰ αἱ ὑπὸ ΕΗΒ, ΒΗΘ δύο ὀρθαῖς ἴσαι εἰσίν· καὶ αἱ ὑπὸ ΒΗΘ, ΗΘΔ ἄρα δύο ὀρθαῖς ἴσαι εἰσίν.

Ἡ ἄρα εἰς τὰς παραλλήλους εὐθείας εὐθεῖα ἐμ-πίπτουσα τάς τε ἐναλλὰξ γωνίας ἴσας ἀλλήλαις 30 ποιεῖ καὶ τὴν ἐκτὸς τῇ ἐντὸς καὶ ἀπεναντίον ἴσην καὶ τὰς ἐντὸς καὶ ἐπὶ τὰ αὐτὰ μέρη δυσὶν ὀρθαῖς ἴσας· ὅπερ ἔδει δεῖξαι.

λ'.

Αἱ τῇ αὐτῇ εὐθείᾳ παράλληλοι καὶ ἀλλήλαις
εἰσὶ παράλληλοι.

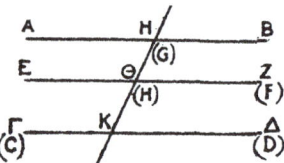

ΕΣΤΩ ἑκατέρα τῶν ΑΒ, ΓΔ τῇ ΕΖ παράλληλος·
λέγω ὅτι καὶ ἡ ΑΒ τῇ ΓΔ ἐστι παράλληλος.

5   Ἐμπιπτέτω γὰρ εἰς αὐτὰς εὐθεῖα ἡ ΗΚ.

Καὶ ἐπεὶ εἰς παραλλήλους εὐθείας τὰς ΑΒ, ΕΖ
εὐθεῖα ἐμπέπτωκεν ἡ ΗΚ, ἴση ἄρα ἡ ὑπὸ ΑΗΚ τῇ
ὑπὸ ΗΘΖ. πάλιν, ἐπεὶ εἰς παραλλήλους εὐθείας τὰς
ΕΖ, ΓΔ εὐθεῖα ἐμπέπτωκεν ἡ ΗΚ, ἴση ἐστὶν ἡ ὑπὸ
10 ΗΘΖ τῇ ὑπὸ ΗΚΔ. ἐδείχθη δὲ καὶ ἡ ὑπὸ ΑΗΚ
τῇ ὑπὸ ΗΘΖ ἴση. καὶ ἡ ὑπὸ ΑΗΚ ἄρα τῇ ὑπὸ
ΗΚΔ ἐστιν ἴση· καί εἰσιν ἐναλλάξ. παράλληλος
ἄρα ἐστὶν ἡ ΑΒ τῇ ΓΔ.

[Αἱ ἄρα τῇ αὐτῇ εὐθείᾳ παράλληλοι καὶ ἀλλήλαις
15 εἰσὶ παράλληλοι·] ὅπερ ἔδει δεῖξαι.

λα΄.

Διὰ τοῦ δοθέντος σημείου τῇ δοθείσῃ εὐθείᾳ
παράλληλον εὐθεῖαν γραμμὴν ἀγαγεῖν.

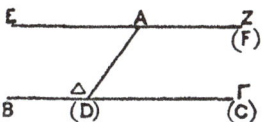

ΕΣΤΩ τὸ μὲν δοθὲν σημεῖον τὸ Α, ἡ δὲ δοθεῖσα
εὐθεῖα ἡ ΒΓ· δεῖ δὴ διὰ τοῦ Α σημείου τῇ ΒΓ
εὐθείᾳ παράλληλον εὐθεῖαν γραμμὴν ἀγαγεῖν. 5
Εἰλήφθω ἐπὶ τῆς ΒΓ τυχὸν σημεῖον τὸ Δ, καὶ
ἐπεζεύχθω ἡ ΑΔ· καὶ συνεστάτω πρὸς τῇ ΔΑ εὐθείᾳ
καὶ τῷ πρὸς αὐτῇ σημείῳ τῷ Α τῇ ὑπὸ ΑΔΓ γωνίᾳ
ἴση ἡ ὑπὸ ΔΑΕ· καὶ ἐκβεβλήσθω ἐπ᾽ εὐθείας τῇ
ΕΑ εὐθεῖα ἡ ΑΖ. 10
Καὶ ἐπεὶ εἰς δύο εὐθείας τὰς ΒΓ, ΕΖ εὐθεῖα ἐμπίπτουσα ἡ ΑΔ τὰς ἐναλλὰξ γωνίας τὰς ὑπὸ ΕΑΔ,
ΑΔΓ ἴσας ἀλλήλαις πεποίηκεν, παράλληλος ἄρα
ἐστὶν ἡ ΕΑΖ τῇ ΒΓ.
Διὰ τοῦ δοθέντος ἄρα σημείου τοῦ Α τῇ δοθείσῃ 15
εὐθείᾳ τῇ ΒΓ παράλληλος εὐθεῖα γραμμὴ ἦκται ἡ
ΕΑΖ· ὅπερ ἔδει ποιῆσαι.

λβ'.

Παντὸς τριγώνου μιᾶς τῶν πλευρῶν προσεκ-
βληθείσης ἡ ἐκτὸς γωνία δυσὶ ταῖς ἐντὸς
καὶ ἀπεναντίον ἴση ἐστίν, καὶ αἱ ἐντὸς τοῦ
τριγώνου τρεῖς γωνίαι δυσὶν ὀρθαῖς ἴσαι εἰσίν.

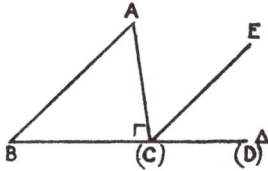

5 Ἔ ΣΤΩ τρίγωνον τὸ ΑΒΓ, καὶ προσεκβεβλήσθω
αὐτοῦ μία πλευρὰ ἡ ΒΓ ἐπὶ τὸ Δ· λέγω ὅτι ἡ
ἐκτὸς γωνία ἡ ὑπὸ ΑΓΔ ἴση ἐστὶ δυσὶ ταῖς ἐντὸς
καὶ ἀπεναντίον ταῖς ὑπὸ ΓΑΒ, ΑΒΓ, καὶ αἱ ἐντὸς
τοῦ τριγώνου τρεῖς γωνίαι αἱ ὑπὸ ΑΒΓ, ΒΓΑ, ΓΑΒ
10 δυσὶν ὀρθαῖς ἴσαι εἰσίν.

Ἤχθω γὰρ διὰ τοῦ Γ σημείου τῇ ΑΒ εὐθείᾳ
παράλληλος ἡ ΓΕ.

Καὶ ἐπεὶ παράλληλός ἐστιν ἡ ΑΒ τῇ ΓΕ, καὶ εἰς
αὐτὰς ἐμπέπτωκεν ἡ ΑΓ, αἱ ἐναλλὰξ γωνίαι αἱ ὑπὸ
15 ΒΑΓ, ΑΓΕ ἴσαι ἀλλήλαις εἰσίν. πάλιν, ἐπεὶ παράλληλός ἐστιν ἡ ΑΒ τῇ ΓΕ, καὶ εἰς αὐτὰς ἐμπέπτωκεν
εὐθεῖα ἡ ΒΔ, ἡ ἐκτὸς γωνία ἡ ὑπὸ ΕΓΔ ἴση ἐστὶ
τῇ ἐντὸς καὶ ἀπεναντίον τῇ ὑπὸ ΑΒΓ. ἐδείχθη δὲ
καὶ ἡ ὑπὸ ΑΓΕ τῇ ὑπὸ ΒΑΓ ἴση· ὅλη ἄρα ἡ ὑπὸ
20 ΑΓΔ γωνία ἴση ἐστὶ δυσὶ ταῖς ἐντὸς καὶ ἀπεναντίον
ταῖς ὑπὸ ΒΑΓ, ΑΒΓ.

Κοινὴ προσκείσθω ἡ ὑπὸ ΑΓΒ· αἱ ἄρα ὑπὸ ΑΓΔ,
ΑΓΒ τρισὶ ταῖς ὑπὸ ΑΒΓ, ΒΓΑ, ΓΑΒ ἴσαι εἰσίν.
ἀλλ' αἱ ὑπὸ ΑΓΔ, ΑΓΒ δυσὶν ὀρθαῖς ἴσαι εἰσίν· καὶ
αἱ ὑπὸ ΑΓΒ, ΓΒΑ, ΓΑΒ ἄρα δυσὶν ὀρθαῖς ἴσαι 25
εἰσίν.

Παντὸς ἄρα τριγώνου μιᾶς τῶν πλευρῶν προσ-
εκβληθείσης ἡ ἐκτὸς γωνία δυσὶ ταῖς ἐντὸς καὶ
ἀπεναντίον ἴση ἐστίν, καὶ αἱ ἐντὸς τοῦ τριγώνου
τρεῖς γωνίαι δυσὶν ὀρθαῖς ἴσαι εἰσίν· ὅπερ ἔδει δεῖξαι. 30

λγ'.

Αἱ τὰς ἴσας τε καὶ παραλλήλους ἐπὶ τὰ αὐτὰ
μέρη ἐπιζευγνύουσαι εὐθεῖαι καὶ αὐταὶ ἴσαι
τε καὶ παράλληλοί εἰσιν.

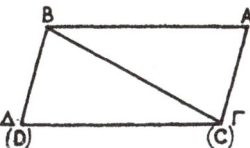

ΕΣΤΩΣΑΝ ἴσαι τε καὶ παράλληλοι αἱ ΑΒ, ΓΔ,
καὶ ἐπιζευγνύτωσαν αὐτὰς ἐπὶ τὰ αὐτὰ μέρη 5
εὐθεῖαι αἱ ΑΓ, ΒΔ· λέγω ὅτι καὶ αἱ ΑΓ, ΒΔ ἴσαι
τε καὶ παράλληλοί εἰσιν.

Ἐπεζεύχθω ἡ ΒΓ. καὶ ἐπεὶ παράλληλός ἐστιν ἡ
ΑΒ τῇ ΓΔ, καὶ εἰς αὐτὰς ἐμπέπτωκεν ἡ ΒΓ, αἱ
ἐναλλὰξ γωνίαι αἱ ὑπὸ ΑΒΓ, ΒΓΔ ἴσαι ἀλλήλαις 10
εἰσίν. καὶ ἐπεὶ ἴση ἐστὶν ἡ ΑΒ τῇ ΓΔ κοινὴ δὲ ἡ
ΒΓ, δύο δὴ αἱ ΑΒ, ΒΓ δύο ταῖς ΒΓ, ΓΔ ἴσαι εἰσίν·
καὶ γωνία ἡ ὑπὸ ΑΒΓ γωνίᾳ τῇ ὑπὸ ΒΓΔ ἴση·

βάσις ἄρα ἡ ΑΓ βάσει τῇ ΒΔ ἐστιν ἴση, καὶ τὸ
15 ΑΒΓ τρίγωνον τῷ ΒΓΔ τριγώνῳ ἴσον ἐστίν, καὶ αἱ
λοιπαὶ γωνίαι ταῖς λοιπαῖς γωνίαις ἴσαι ἔσονται ἑκατέρα ἑκατέρᾳ, ὑφ' ἃς αἱ ἴσαι πλευραὶ ὑποτείνουσιν·
ἴση ἄρα ἡ ὑπὸ ΑΓΒ γωνία τῇ ὑπὸ ΓΒΔ. καὶ ἐπεὶ
εἰς δύο εὐθείας τὰς ΑΓ, ΒΔ εὐθεῖα ἐμπίπτουσα ἡ
20 ΒΓ τὰς ἐναλλὰξ γωνίας ἴσας ἀλλήλαις πεποίηκεν,
παράλληλος ἄρα ἐστὶν ἡ ΑΓ τῇ ΒΔ. ἐδείχθη δὲ
αὐτῇ καὶ ἴση.
Αἱ ἄρα τὰς ἴσας τε καὶ παραλλήλους ἐπὶ τὰ αὐτὰ
μέρη ἐπιζευγνύουσαι εὐθεῖαι καὶ αὐταὶ ἴσαι τε καὶ
25 παράλληλοί εἰσιν· ὅπερ ἔδει δεῖξαι.

λδʹ.

Τῶν παραλληλογράμμων χωρίων αἱ ἀπεναντίον
πλευραί τε καὶ γωνίαι ἴσαι ἀλλήλαις εἰσίν,
καὶ ἡ διάμετρος αὐτὰ δίχα τέμνει.

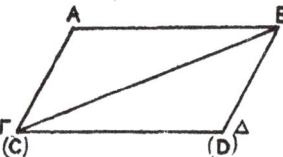

ΕΣΤΩ παραλληλόγραμμον χωρίον τὸ ΑΓΔΒ, διά-
5 μετρος δὲ αὐτοῦ ἡ ΒΓ· λέγω ὅτι τοῦ ΑΓΔΒ
παραλληλογράμμου αἱ ἀπεναντίον πλευραί τε καὶ
γωνίαι ἴσαι ἀλλήλαις εἰσίν, καὶ ἡ ΒΓ διάμετρος
αὐτὸ δίχα τέμνει.
Ἐπεὶ γὰρ παράλληλός ἐστιν ἡ ΑΒ τῇ ΓΔ, καὶ

εἰς αὐτὰς ἐμπέπτωκεν εὐθεῖα ἡ ΒΓ, αἱ ἐναλλὰξ γω- 10
νίαι αἱ ὑπὸ ΑΒΓ, ΒΓΔ ἴσαι ἀλλήλαις εἰσίν. πάλιν
ἐπεὶ παράλληλός ἐστιν ἡ ΑΓ τῇ ΒΔ, καὶ εἰς αὐτὰς
ἐμπέπτωκεν ἡ ΒΓ, αἱ ἐναλλὰξ γωνίαι αἱ ὑπὸ ΑΓΒ,
ΓΒΔ ἴσαι ἀλλήλαις εἰσίν. δύο δὴ τρίγωνά ἐστι τὰ
ΑΒΓ, ΒΓΔ τὰς δύο γωνίας τὰς ὑπὸ ΑΒΓ, ΒΓΑ 15
δυσὶ ταῖς ὑπὸ ΒΓΔ, ΓΒΔ ἴσας ἔχοντα ἑκατέραν
ἑκατέρᾳ καὶ μίαν πλευρὰν μιᾷ πλευρᾷ ἴσην τὴν πρὸς
ταῖς ἴσαις γωνίαις κοινὴν αὐτῶν τὴν ΒΓ· καὶ τὰς
λοιπὰς ἄρα πλευρὰς ταῖς λοιπαῖς ἴσας ἕξει ἑκατέραν
ἑκατέρᾳ καὶ τὴν λοιπὴν γωνίαν τῇ λοιπῇ γωνίᾳ· ἴση 20
ἄρα ἡ μὲν ΑΒ πλευρὰ τῇ ΓΔ, ἡ δὲ ΑΓ τῇ ΒΔ, καὶ
ἔτι ἴση ἐστὶν ἡ ὑπὸ ΒΑΓ γωνία τῇ ὑπὸ ΓΔΒ. καὶ
ἐπεὶ ἴση ἐστὶν ἡ μὲν ὑπὸ ΑΒΓ γωνία τῇ ὑπὸ ΒΓΔ,
ἡ δὲ ὑπὸ ΓΒΔ τῇ ὑπὸ ΑΓΒ, ὅλη ἄρα ἡ ὑπὸ ΑΒΔ
ὅλη τῇ ὑπὸ ΑΓΔ ἐστιν ἴση. ἐδείχθη δὲ καὶ ἡ ὑπὸ 25
ΒΑΓ τῇ ὑπὸ ΓΔΒ ἴση.

Τῶν ἄρα παραλληλογράμμων χωρίων αἱ ἀπεναντίον πλευραί τε καὶ γωνίαι ἴσαι ἀλλήλαις εἰσίν.

Λέγω δὴ ὅτι καὶ ἡ διάμετρος αὐτὰ δίχα τέμνει.
ἐπεὶ γὰρ ἴση ἐστὶν ἡ ΑΒ τῇ ΓΔ, κοινὴ δὲ ἡ ΒΓ, 30
δύο δὴ αἱ ΑΒ, ΒΓ δυσὶ ταῖς ΓΔ, ΒΓ ἴσαι εἰσὶν
ἑκατέρα ἑκατέρᾳ· καὶ γωνία ἡ ὑπὸ ΑΒΓ γωνίᾳ τῇ
ὑπὸ ΒΓΔ ἴση. καὶ βάσις ἄρα ἡ ΑΓ τῇ ΔΒ ἴση.
καὶ τὸ ΑΒΓ [ἄρα] τρίγωνον τῷ ΒΓΔ τριγώνῳ ἴσον
ἐστίν. 35

Ἡ ἄρα ΒΓ διάμετρος δίχα τέμνει τὸ ΑΒΓΔ
παραλληλόγραμμον· ὅπερ ἔδει δεῖξαι.

λε'.

Τὰ παραλληλόγραμμα τὰ ἐπὶ τῆς αὐτῆς βάσεως
ὄντα καὶ ἐν ταῖς αὐταῖς παραλλήλοις ἴσα
ἀλλήλοις ἐστίν.

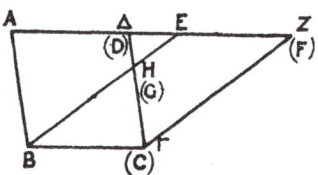

ΕΣΤΩ παραλληλόγραμμα τὰ ΑΒΓΔ, ΕΒΓΖ ἐπὶ
5   τῆς αὐτῆς βάσεως τῆς ΒΓ καὶ ἐν ταῖς αὐταῖς
παραλλήλοις ταῖς ΑΖ, ΒΓ· λέγω ὅτι ἴσον ἐστὶ τὸ
ΑΒΓΔ τῷ ΕΒΓΖ παραλληλογράμμῳ.

Ἐπεὶ γὰρ παραλληλόγραμμόν ἐστι τὸ ΑΒΓΔ, ἴση
ἐστὶν ἡ ΑΔ τῇ ΒΓ. διὰ τὰ αὐτὰ δὴ καὶ ἡ ΕΖ τῇ
10   ΒΓ ἐστιν ἴση· ὥστε καὶ ἡ ΑΔ τῇ ΕΖ ἐστιν ἴση· καὶ
κοινὴ ἡ ΔΕ· ὅλη ἄρα ἡ ΑΕ ὅλῃ τῇ ΔΖ ἐστιν ἴση.
ἔστι δὲ καὶ ἡ ΑΒ τῇ ΔΓ ἴση· δύο δὴ αἱ ΕΑ, ΑΒ
δύο ταῖς ΖΔ, ΔΓ ἴσαι εἰσὶν ἑκατέρα ἑκατέρᾳ· καὶ
γωνία ἡ ὑπὸ ΖΔΓ γωνίᾳ τῇ ὑπὸ ΕΑΒ ἐστιν ἴση ἡ
15   ἐκτὸς τῇ ἐντός· βάσις ἄρα ἡ ΕΒ βάσει τῇ ΖΓ ἴση
ἐστίν, καὶ τὸ ΕΑΒ τρίγωνον τῷ ΔΖΓ τριγώνῳ ἴσον
ἔσται· κοινὸν ἀφῃρήσθω τὸ ΔΗΕ· λοιπὸν ἄρα τὸ
ΑΒΗΔ τραπέζιον λοιπῷ τῷ ΕΗΓΖ τραπεζίῳ ἐστὶν
ἴσον· κοινὸν προσκείσθω τὸ ΗΒΓ τρίγωνον· ὅλον
20   ἄρα τὸ ΑΒΓΔ παραλληλόγραμμον ὅλῳ τῷ ΕΒΓΖ
παραλληλογράμμῳ ἴσον ἐστίν.

Τὰ ἄρα παραλληλόγραμμα τὰ ἐπὶ τῆς αὐτῆς βάσεως ὄντα καὶ ἐν ταῖς αὐταῖς παραλλήλοις ἴσα ἀλλήλοις ἐστίν· ὅπερ ἔδει δεῖξαι.

λϛ'.

Τὰ παραλληλόγραμμα τὰ ἐπὶ ἴσων βάσεων ὄντα καὶ ἐν ταῖς αὐταῖς παραλλήλοις ἴσα ἀλλήλοις ἐστίν.

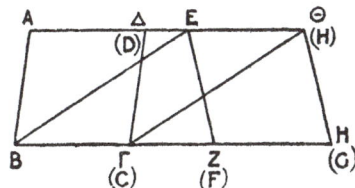

ΕΣΤΩ παραλληλόγραμμα τὰ ΑΒΓΔ, ΕΖΗΘ ἐπὶ ἴσων βάσεων ὄντα τῶν ΒΓ, ΖΗ καὶ ἐν ταῖς 5 αὐταῖς παραλλήλοις ταῖς ΑΘ, ΒΗ· λέγω ὅτι ἴσον ἐστὶ τὸ ΑΒΓΔ παραλληλόγραμμον τῷ ΕΖΗΘ.

Ἐπεζεύχθωσαν γὰρ αἱ ΒΕ, ΓΘ. καὶ ἐπεὶ ἴση ἐστὶν ἡ ΒΓ τῇ ΖΗ, ἀλλὰ ἡ ΖΗ τῇ ΕΘ ἐστιν ἴση, καὶ ἡ ΒΓ ἄρα τῇ ΕΘ ἐστιν ἴση. εἰσὶ δὲ καὶ παράλ- 10 ληλοι. καὶ ἐπιζευγνύουσιν αὐτὰς αἱ ΕΒ, ΘΓ· αἱ δὲ τὰς ἴσας τε καὶ παραλλήλους ἐπὶ τὰ αὐτὰ μέρη ἐπιζευγνύουσαι ἴσαι τε καὶ παράλληλοί εἰσι [καὶ αἱ ΕΒ, ΘΓ ἄρα ἴσαι τέ εἰσι καὶ παράλληλοι]. παραλληλόγραμμον ἄρα ἐστὶ τὸ ΕΒΓΘ. καί ἐστιν ἴσον 15 τῷ ΑΒΓΔ· βάσιν τε γὰρ αὐτῷ τὴν αὐτὴν ἔχει τὴν ΒΓ, καὶ ἐν ταῖς αὐταῖς παραλλήλοις ἐστὶν αὐτῷ ταῖς ΒΓ, ΑΘ. διὰ τὰ αὐτὰ δὴ καὶ τὸ ΕΖΗΘ τῷ αὐτῷ

τῷ ΕΒΓΘ ἐστιν ἴσον· ὥστε καὶ τὸ ΑΒΓΔ παραλ-
20 ληλόγραμμον τῷ ΕΖΗΘ ἐστιν ἴσον.

Τὰ ἄρα παραλληλόγραμμα τὰ ἐπὶ ἴσων βάσεων
ὄντα καὶ ἐν ταῖς αὐταῖς παραλλήλοις ἴσα ἀλλήλοις
ἐστίν· ὅπερ ἔδει δεῖξαι.

λζ'.

Τὰ τρίγωνα τὰ ἐπὶ τῆς αὐτῆς βάσεως ὄντα καὶ
ἐν ταῖς αὐταῖς παραλλήλοις ἴσα ἀλλήλοις
ἐστίν.

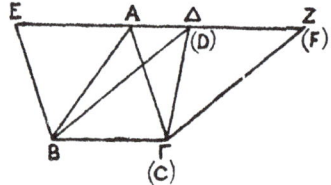

Ε῎ΣΤΩ τρίγωνα τὰ ΑΒΓ, ΔΒΓ ἐπὶ τῆς αὐτῆς
5   βάσεως τῆς ΒΓ καὶ ἐν ταῖς αὐταῖς παραλλήλοις
ταῖς ΑΔ, ΒΓ· λέγω ὅτι ἴσον ἐστὶ τὸ ΑΒΓ τρίγωνον
τῷ ΔΒΓ τριγώνῳ.

Ἐκβεβλήσθω ἡ ΑΔ ἐφ' ἑκάτερα τὰ μέρη ἐπὶ τὰ
Ε, Ζ, καὶ διὰ μὲν τοῦ Β τῇ ΓΑ παράλληλος ἤχθω
10 ἡ ΒΕ, διὰ δὲ τοῦ Γ τῇ ΒΔ παράλληλος ἤχθω ἡ ΓΖ.
παραλληλόγραμμον ἄρα ἐστὶν ἑκάτερον τῶν ΕΒΓΑ,
ΔΒΓΖ· καί εἰσιν ἴσα· ἐπί τε γὰρ τῆς αὐτῆς βάσεώς
εἰσι τῆς ΒΓ καὶ ἐν ταῖς αὐταῖς παραλλήλοις ταῖς
ΒΓ, ΕΖ· καί ἐστι τοῦ μὲν ΕΒΓΑ παραλληλογράμ-
15 μου ἥμισυ τὸ ΑΒΓ τρίγωνον· ἡ γὰρ ΑΒ διάμετρος
αὐτὸ δίχα τέμνει· τοῦ δὲ ΔΒΓΖ παραλληλογράμμου

ἥμισυ τὸ ΔΒΓ τρίγωνον· ἡ γὰρ ΔΓ διάμετρος αὐτὸ
δίχα τέμνει. [τὰ δὲ τῶν ἴσων ἡμίση ἴσα ἀλλήλοις
ἐστίν]. ἴσον ἄρα ἐστὶ τὸ ΑΒΓ τρίγωνον τῷ ΔΒΓ
τριγώνῳ.   20
Τὰ ἄρα τρίγωνα τὰ ἐπὶ τῆς αὐτῆς βάσεως ὄντα
καὶ ἐν ταῖς αὐταῖς παραλλήλοις ἴσα ἀλλήλοις ἐστίν·
ὅπερ ἔδει δεῖξαι.

λη'.

Τὰ τρίγωνα τὰ ἐπὶ ἴσων βάσεων ὄντα καὶ ἐν
ταῖς αὐταῖς παραλλήλοις ἴσα ἀλλήλοις ἐστίν.

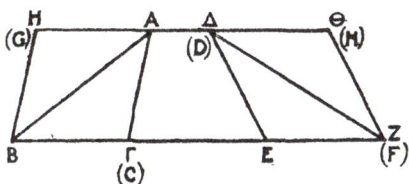

Ἔστω τρίγωνα τὰ ΑΒΓ, ΔΕΖ ἐπὶ ἴσων βάσεων
τῶν ΒΓ, ΕΖ καὶ ἐν ταῖς αὐταῖς παραλλήλοις
ταῖς ΒΖ, ΑΔ· λέγω ὅτι ἴσον ἐστὶ τὸ ΑΒΓ τρίγωνον   5
τῷ ΔΕΖ τριγώνῳ.
Ἐκβεβλήσθω γὰρ ἡ ΑΔ ἐφ' ἑκάτερα τὰ μέρη ἐπὶ
τὰ Η, Θ, καὶ διὰ μὲν τοῦ Β τῇ ΓΑ παράλληλος
ἤχθω ἡ ΒΗ, διὰ δὲ τοῦ Ζ τῇ ΔΕ παράλληλος ἤχθω
ἡ ΖΘ.   παραλληλόγραμμον ἄρα ἐστὶν ἑκάτερον τῶν   10
ΗΒΓΑ, ΔΕΖΘ· καὶ ἴσον τὸ ΗΒΓΑ τῷ ΔΕΖΘ· ἐπί
τε γὰρ ἴσων βάσεών εἰσι τῶν ΒΓ, ΕΖ καὶ ἐν ταῖς
αὐταῖς παραλλήλοις ταῖς ΒΖ, ΗΘ· καί ἐστι τοῦ μὲν
ΗΒΓΑ παραλληλογράμμου ἥμισυ τὸ ΑΒΓ τρίγωνον·

¹⁵ ἡ γὰρ ΑΒ διάμετρος αὐτὸ δίχα τέμνει· τοῦ δὲ ΔΕΖΘ
παραλληλογράμμου ἥμισυ τὸ ΖΕΔ τρίγωνον· ἡ γὰρ
ΔΖ διάμετρος αὐτὸ δίχα τέμνει [τὰ δὲ τῶν ἴσων
ἡμίση ἴσα ἀλλήλοις ἐστίν]. ἴσον ἄρα ἐστὶ τὸ ΑΒΓ
τρίγωνον τῷ ΔΕΖ τριγώνῳ.

²⁰   Τὰ ἄρα τρίγωνα τὰ ἐπὶ ἴσων βάσεων ὄντα καὶ ἐν
ταῖς αὐταῖς παραλλήλοις ἴσα ἀλλήλοις ἐστίν· ὅπερ
ἔδει δεῖξαι.

λθ'.

Τὰ ἴσα τρίγωνα τὰ ἐπὶ τῆς αὐτῆς βάσεως ὄντα
καὶ ἐπὶ τὰ αὐτὰ μέρη καὶ ἐν ταῖς αὐταῖς
παραλλήλοις ἐστίν.

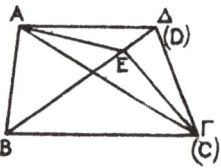

ἘΣΤΩ ἴσα τρίγωνα τὰ ΑΒΓ, ΔΒΓ ἐπὶ τῆς αὐτῆς
⁵   βάσεως ὄντα καὶ ἐπὶ τὰ αὐτὰ μέρη τῆς ΒΓ·
λέγω ὅτι καὶ ἐν ταῖς αὐταῖς παραλλήλοις ἐστίν.
Ἐπεζεύχθω γὰρ ἡ ΑΔ· λέγω ὅτι παράλληλός
ἐστιν ἡ ΑΔ τῇ ΒΓ.
Εἰ γὰρ μή, ἤχθω διὰ τοῦ Α σημείου τῇ ΒΓ εὐθείᾳ
¹⁰ παράλληλος ἡ ΑΕ, καὶ ἐπεζεύχθω ἡ ΕΓ. ἴσον ἄρα
ἐστὶ τὸ ΑΒΓ τρίγωνον τῷ ΕΒΓ τριγώνῳ· ἐπί τε γὰρ
τῆς αὐτῆς βάσεώς ἐστιν αὐτῷ τῆς ΒΓ καὶ ἐν ταῖς
αὐταῖς παραλλήλοις. ἀλλὰ τὸ ΑΒΓ τῷ ΔΒΓ ἐστιν
ἴσον· καὶ τὸ ΔΒΓ ἄρα τῷ ΕΒΓ ἴσον ἐστὶ τὸ μεῖζον

τῷ ἐλάσσονι· ὅπερ ἐστὶν ἀδύνατον· οὐκ ἄρα παράλ- 15
ληλός ἐστιν ἡ ΑΕ τῇ ΒΓ. ὁμοίως δὴ δείξομεν ὅτι
οὐδ' ἄλλη τις πλὴν τῆς ΑΔ· ἡ ΑΔ ἄρα τῇ ΒΓ ἐστι
παράλληλος.

Τὰ ἄρα ἴσα τρίγωνα τὰ ἐπὶ τῆς αὐτῆς βάσεως
ὄντα καὶ ἐπὶ τὰ αὐτὰ μέρη καὶ ἐν ταῖς αὐταῖς παραλ- 20
λήλοις ἐστίν· ὅπερ ἔδει δεῖξαι.

μ'.

Τὰ ἴσα τρίγωνα τὰ ἐπὶ ἴσων βάσεων ὄντα καὶ
ἐπὶ τὰ αὐτὰ μέρη καὶ ἐν ταῖς αὐταῖς παραλλήλοις ἐστίν.

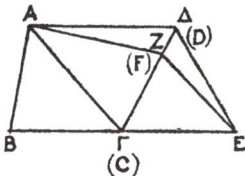

ΕΣΤΩ ἴσα τρίγωνα τὰ ΑΒΓ, ΓΔΕ ἐπὶ ἴσων
βάσεων τῶν ΒΓ, ΓΕ καὶ ἐπὶ τὰ αὐτὰ μέρη. 5
λέγω ὅτι καὶ ἐν ταῖς αὐταῖς παραλλήλοις ἐστίν.

Ἐπεζεύχθω γὰρ ἡ ΑΔ· λέγω ὅτι παράλληλός
ἐστιν ἡ ΑΔ τῇ ΒΕ.

Εἰ γὰρ μή, ἤχθω διὰ τοῦ Α τῇ ΒΕ παράλληλος
ἡ ΑΖ, καὶ ἐπεζεύχθω ἡ ΖΕ. ἴσον ἄρα ἐστὶ τὸ ΑΒΓ 10
τρίγωνον τῷ ΖΓΕ τριγώνῳ· ἐπί τε γὰρ ἴσων βάσεών
εἰσι τῶν ΒΓ, ΓΕ καὶ ἐν ταῖς αὐταῖς παραλλήλοις
ταῖς ΒΕ, ΑΖ. ἀλλὰ τὸ ΑΒΓ τρίγωνον ἴσον ἐστὶ τῷ
ΔΓΕ [τριγώνῳ]· καὶ τὸ ΔΓΕ ἄρα [τρίγωνον] ἴσον

15 ἐστὶ τῷ ΖΓΕ τριγώνῳ τὸ μεῖζον τῷ ἐλάσσονι· ὅπερ
ἐστὶν ἀδύνατον· οὐκ ἄρα παράλληλος ἡ ΑΖ τῇ ΒΕ.
ὁμοίως δὴ δείξομεν ὅτι οὐδ᾽ ἄλλη τις πλὴν τῆς ΑΔ·
ἡ ΑΔ ἄρα τῇ ΒΕ ἐστι παράλληλος.

Τὰ ἄρα ἴσα τρίγωνα τὰ ἐπὶ ἴσων βάσεων ὄντα
20 καὶ ἐπὶ τὰ αὐτὰ μέρη καὶ ἐν ταῖς αὐταῖς παραλλήλοις
ἐστίν· ὅπερ ἔδει δεῖξαι.

μα΄.

Ἐὰν παραλληλόγραμμον τριγώνῳ βάσιν τε ἔχῃ
τὴν αὐτὴν καὶ ἐν ταῖς αὐταῖς παραλλήλοις ᾖ,
διπλάσιόν ἐστι τὸ παραλληλόγραμμον τοῦ
τριγώνου.

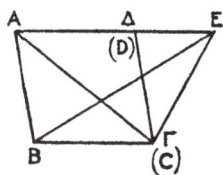

5 ΠΑΡΑΛΛΗΛΟΓΡΑΜΜΟΝ γὰρ τὸ ΑΒΓΔ τριγώνῳ τῷ ΕΒΓ βάσιν τε ἐχέτω τὴν αὐτὴν τὴν
ΒΓ καὶ ἐν ταῖς αὐταῖς παραλλήλοις ἔστω ταῖς ΒΓ,
ΑΕ· λέγω ὅτι διπλάσιόν ἐστι τὸ ΑΒΓΔ παραλληλόγραμμον τοῦ ΒΕΓ τριγώνου.

10 Ἐπεζεύχθω γὰρ ἡ ΑΓ. ἴσον δή ἐστι τὸ ΑΒΓ
τρίγωνον τῷ ΕΒΓ τριγώνῳ· ἐπί τε γὰρ τῆς αὐτῆς
βάσεώς ἐστιν αὐτῷ τῆς ΒΓ καὶ ἐν ταῖς αὐταῖς
παραλλήλοις ταῖς ΒΓ, ΑΕ. ἀλλὰ τὸ ΑΒΓΔ παραλληλόγραμμον διπλάσιόν ἐστι τοῦ ΑΒΓ τριγώνου· ἡ

γὰρ ΑΓ διάμετρος αὐτὸ δίχα τέμνει· ὥστε τὸ ΑΒΓΔ 15
παραλληλόγραμμον καὶ τοῦ ΕΒΓ τριγώνου ἐστὶ
διπλάσιον.

Ἐὰν ἄρα παραλληλόγραμμον τριγώνῳ βάσιν τε
ἔχῃ τὴν αὐτὴν καὶ ἐν ταῖς αὐταῖς παραλλήλοις ᾖ,
διπλάσιόν ἐστι τὸ παραλληλόγραμμον τοῦ τριγώνου· 20
ὅπερ ἔδει δεῖξαι.

μβ'.

Τῷ δοθέντι τριγώνῳ ἴσον παραλληλόγραμμον
συστήσασθαι ἐν τῇ δοθείσῃ γωνίᾳ εὐθυγράμμῳ.

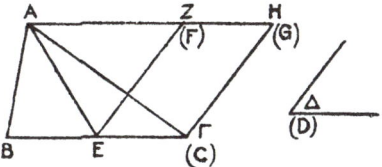

ΕΣΤΩ τὸ μὲν δοθὲν τρίγωνον τὸ ΑΒΓ, ἡ δὲ δοθεῖσα
γωνία εὐθύγραμμος ἡ Δ· δεῖ δὴ τῷ ΑΒΓ τρι- 5
γώνῳ ἴσον παραλληλόγραμμον συστήσασθαι ἐν τῇ
Δ γωνίᾳ εὐθυγράμμῳ.

Τετμήσθω ἡ ΒΓ δίχα κατὰ τὸ Ε, καὶ ἐπεζεύχθω
ἡ ΑΕ, καὶ συνεστάτω πρὸς τῇ ΕΓ εὐθείᾳ καὶ τῷ
πρὸς αὐτῇ σημείῳ τῷ Ε τῇ Δ γωνίᾳ ἴση ἡ ὑπὸ 10
ΓΕΖ, καὶ διὰ μὲν τοῦ Α τῇ ΕΓ παράλληλος ἤχθω ἡ
ΑΗ, διὰ δὲ τοῦ Γ τῇ ΕΖ παράλληλος ἤχθω ἡ ΓΗ·
παραλληλόγραμμον ἄρα ἐστὶ τὸ ΖΕΓΗ. καὶ ἐπεὶ
ἴση ἐστὶν ἡ ΒΕ τῇ ΕΓ, ἴσον ἐστὶ καὶ τὸ ΑΒΕ
τρίγωνον τῷ ΑΕΓ τριγώνῳ· ἐπί τε γὰρ ἴσων βάσεών 15

εἰσι τῶν ΒΕ, ΕΓ καὶ ἐν ταῖς αὐταῖς παραλλήλοις
ταῖς ΒΓ, ΑΗ· διπλάσιον ἄρα ἐστὶ τὸ ΑΒΓ τρίγωνον
τοῦ ΑΕΓ τριγώνου. ἔστι δὲ καὶ τὸ ΖΕΓΗ παραλληλόγραμμον διπλάσιον τοῦ ΑΕΓ τριγώνου· βάσιν
20 τε γὰρ αὐτῷ τὴν αὐτὴν ἔχει καὶ ἐν ταῖς αὐταῖς ἐστιν
αὐτῷ παραλλήλοις· ἴσον ἄρα ἐστὶ τὸ ΖΕΓΗ παραλληλόγραμμον τῷ ΑΒΓ τριγώνῳ. καὶ ἔχει τὴν ὑπὸ
ΓΕΖ γωνίαν ἴσην τῇ δοθείσῃ τῇ Δ.
Τῷ ἄρα δοθέντι τριγώνῳ τῷ ΑΒΓ ἴσον παραλ-
25 ληλόγραμμον συνέσταται τὸ ΖΕΓΗ ἐν γωνίᾳ τῇ
ὑπὸ ΓΕΖ, ἥτις ἐστὶν ἴση τῇ Δ· ὅπερ ἔδει ποιῆσαι.

μγ΄.

Παντὸς παραλληλογράμμου τῶν περὶ τὴν διάμετρον παραλληλογράμμων τὰ παραπληρώματα ἴσα ἀλλήλοις ἐστίν.

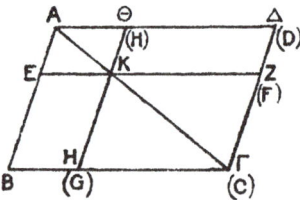

5 ΕΣΤΩ παραλληλόγραμμον τὸ ΑΒΓΔ, διάμετρος
δὲ αὐτοῦ ἡ ΑΓ, περὶ δὲ τὴν ΑΓ παραλληλόγραμμα μὲν ἔστω τὰ ΕΘ, ΖΗ, τὰ δὲ λεγόμενα
παραπληρώματα τὰ ΒΚ, ΚΔ· λέγω ὅτι ἴσον ἐστὶ
τὸ ΒΚ παραπλήρωμα τῷ ΚΔ παραπληρώματι.

Ἐπεὶ γὰρ παραλληλόγραμμόν ἐστι τὸ ΑΒΓΔ, διάμετρος δὲ αὐτοῦ ἡ ΑΓ, ἴσον ἐστὶ τὸ ΑΒΓ τρίγωνον 10 τῷ ΑΓΔ τριγώνῳ. πάλιν, ἐπεὶ παραλληλόγραμμόν ἐστι τὸ ΕΘ, διάμετρος δὲ αὐτοῦ ἐστιν ἡ ΑΚ, ἴσον ἐστὶ τὸ ΑΕΚ τρίγωνον τῷ ΑΘΚ τριγώνῳ. διὰ τὰ αὐτὰ δὴ καὶ τὸ ΚΖΓ τρίγωνον τῷ ΚΗΓ ἐστιν ἴσον. ἐπεὶ οὖν τὸ μὲν ΑΕΚ τρίγωνον τῷ ΑΘΚ τριγώνῳ 15 ἐστὶν ἴσον, τὸ δὲ ΚΖΓ τῷ ΚΗΓ, τὸ ΑΕΚ τρίγωνον μετὰ τοῦ ΚΗΓ ἴσον ἐστὶ τῷ ΑΘΚ τριγώνῳ μετὰ τοῦ ΚΖΓ· ἔστι δὲ καὶ ὅλον τὸ ΑΒΓ τρίγωνον ὅλῳ τῷ ΑΔΓ ἴσον· λοιπὸν ἄρα τὸ ΒΚ παραπλήρωμα λοιπῷ τῷ ΚΔ παραπληρώματί ἐστιν ἴσον. 20

Παντὸς ἄρα παραλληλογράμμου χωρίου τῶν περὶ τὴν διάμετρον παραλληλογράμμων τὰ παραπληρώματα ἴσα ἀλλήλοις ἐστίν· ὅπερ ἔδει δεῖξαι.

μδ'.

Παρὰ τὴν δοθεῖσαν εὐθεῖαν τῷ δοθέντι τριγώνῳ
ἴσον παραλληλόγραμμον παραβαλεῖν ἐν τῇ
δοθείσῃ γωνίᾳ εὐθυγράμμῳ.

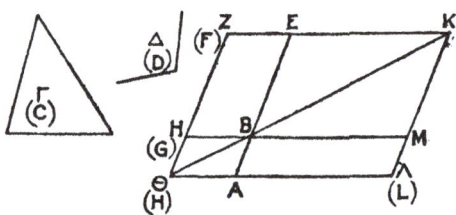

5   ΕΣΤΩ ἡ μὲν δοθεῖσα εὐθεῖα ἡ ΑΒ, τὸ δὲ δοθὲν
τρίγωνον τὸ Γ, ἡ δὲ δοθεῖσα γωνία εὐθύγραμμος
ἡ Δ· δεῖ δὴ παρὰ τὴν δοθεῖσαν εὐθεῖαν τὴν ΑΒ τῷ
δοθέντι τριγώνῳ τῷ Γ ἴσον παραλληλόγραμμον παραβαλεῖν ἐν ἴσῃ τῇ Δ γωνίᾳ.

Συνεστάτω τῷ Γ τριγώνῳ ἴσον παραλληλόγραμμον
10   τὸ ΒΕΖΗ ἐν γωνίᾳ τῇ ὑπὸ ΕΒΗ, ἥ ἐστιν ἴση τῇ Δ·
καὶ κείσθω ὥστε ἐπ' εὐθείας εἶναι τὴν ΒΕ τῇ ΑΒ,
καὶ διήχθω ἡ ΖΗ ἐπὶ τὸ Θ, καὶ διὰ τοῦ Α ὁποτέρᾳ
τῶν ΒΗ, ΕΖ παράλληλος ἤχθω ἡ ΑΘ, καὶ ἐπεζεύχθω
ἡ ΘΒ. καὶ ἐπεὶ εἰς παραλλήλους τὰς ΑΘ, ΕΖ εὐθεῖα
15   ἐνέπεσεν ἡ ΘΖ, αἱ ἄρα ὑπὸ ΑΘΖ, ΘΖΕ γωνίαι δυσὶν
ὀρθαῖς εἰσιν ἴσαι. αἱ ἄρα ὑπὸ ΒΘΗ, ΗΖΕ δύο
ὀρθῶν ἐλάσσονές εἰσιν· αἱ δὲ ἀπὸ ἐλασσόνων ἢ δύο
ὀρθῶν εἰς ἄπειρον ἐκβαλλόμεναι συμπίπτουσιν· αἱ

ΘΒ, ΖΕ ἄρα ἐκβαλλόμεναι συμπεσοῦνται. ἐκβεβλήσθωσαν καὶ συμπιπτέτωσαν κατὰ τὸ Κ, καὶ διὰ τοῦ 20
Κ σημείου ὁποτέρᾳ τῶν ΕΑ, ΖΘ παράλληλος ἤχθω
ἡ ΚΛ, καὶ ἐκβεβλήσθωσαν αἱ ΘΑ, ΗΒ ἐπὶ τὰ Λ,
Μ σημεῖα. παραλληλόγραμμον ἄρα ἐστὶ τὸ ΘΛΚΖ,
διάμετρος δὲ αὐτοῦ ἡ ΘΚ, περὶ δὲ τὴν ΘΚ παραλληλόγραμμα μὲν τὰ ΑΗ, ΜΕ, τὰ δὲ λεγόμενα 25
παραπληρώματα τὰ ΛΒ, ΒΖ· ἴσον ἄρα ἐστὶ τὸ ΛΒ
τῷ ΒΖ. ἀλλὰ τὸ ΒΖ τῷ Γ τριγώνῳ ἐστὶν ἴσον· καὶ
τὸ ΛΒ ἄρα τῷ Γ ἐστιν ἴσον. καὶ ἐπεὶ ἴση ἐστὶν ἡ
ὑπὸ ΗΒΕ γωνία τῇ ὑπὸ ΑΒΜ, ἀλλὰ ἡ ὑπὸ ΗΒΕ
τῇ Δ ἐστιν ἴση, καὶ ἡ ὑπὸ ΑΒΜ ἄρα τῇ Δ γωνίᾳ 30
ἐστὶν ἴση.

Παρὰ τὴν δοθεῖσαν ἄρα εὐθεῖαν τὴν ΑΒ τῷ δοθέντι
τριγώνῳ τῷ Γ ἴσον παραλληλόγραμμον παραβέβλη-
ται τὸ ΛΒ ἐν γωνίᾳ τῇ ὑπὸ ΑΒΜ, ἥ ἐστιν ἴση τῇ
Δ· ὅπερ ἔδει ποιῆσαι.   35

με΄.

Τῷ δοθέντι εὐθυγράμμῳ ἴσον παραλληλόγραμμον συστήσασθαι ἐν τῇ δοθείσῃ γωνίᾳ εὐθυγράμμῳ.

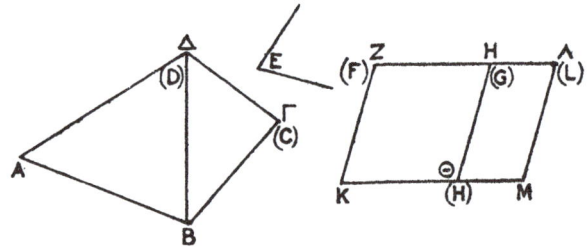

5 ΕΣΤΩ τὸ μὲν δοθὲν εὐθύγραμμον τὸ ΑΒΓΔ, ἡ δὲ δοθεῖσα γωνία εὐθύγραμμος ἡ Ε· δεῖ δὴ τῷ ΑΒΓΔ εὐθυγράμμῳ ἴσον παραλληλόγραμμον συστήσασθαι ἐν τῇ δοθείσῃ γωνίᾳ τῇ Ε.

Ἐπεζεύχθω ἡ ΔΒ, καὶ συνεστάτω τῷ ΑΒΔ τριγώνῳ ἴσον παραλληλόγραμμον τὸ ΖΘ ἐν τῇ ὑπὸ ΘΚΖ 10 γωνίᾳ, ἥ ἐστιν ἴση τῇ Ε· καὶ παραβεβλήσθω παρὰ τὴν ΗΘ εὐθεῖαν τῷ ΔΒΓ τριγώνῳ ἴσον παραλληλόγραμμον τὸ ΗΜ ἐν τῇ ὑπὸ ΗΘΜ γωνίᾳ, ἥ ἐστιν ἴση τῇ Ε. καὶ ἐπεὶ ἡ Ε γωνία ἑκατέρᾳ τῶν ὑπὸ ΘΚΖ, ΗΘΜ· ἐστιν ἴση, καὶ ἡ ὑπὸ ΘΚΖ ἄρα τῇ ὑπὸ 15 ΗΘΜ ἐστιν ἴση. κοινὴ προσκείσθω ἡ ὑπὸ ΚΘΗ· αἱ ἄρα ὑπὸ ΖΚΘ, ΚΘΗ ταῖς ὑπὸ ΚΘΗ, ΗΘΜ ἴσαι εἰσίν. ἀλλ᾽ αἱ ὑπὸ ΖΚΘ, ΚΘΗ δυσὶν ὀρθαῖς ἴσαι εἰσίν· καὶ αἱ ὑπὸ ΚΘΗ, ΗΘΜ ἄρα δύο ὀρθαῖς ἴσαι

εἰσίν. πρὸς δή τινι εὐθείᾳ τῇ ΗΘ καὶ τῷ πρὸς αὐτῇ
σημείῳ τῷ Θ δύο εὐθεῖαι αἱ ΚΘ, ΘΜ μὴ ἐπὶ τὰ 20
αὐτὰ μέρη κείμεναι τὰς ἐφεξῆς γωνίας δύο ὀρθαῖς
ἴσας ποιοῦσιν· ἐπ᾿ εὐθείας ἄρα ἐστὶν ἡ ΚΘ τῇ ΘΜ.
καὶ ἐπεὶ εἰς παραλλήλους τὰς ΚΜ, ΖΗ εὐθεῖα ἐνέπεσεν ἡ ΘΗ, αἱ ἐναλλὰξ γωνίαι αἱ ὑπὸ ΜΘΗ, ΘΗΖ
ἴσαι ἀλλήλαις εἰσίν. κοινὴ προσκείσθω ἡ ὑπὸ ΘΗΛ· 25
αἱ ἄρα ὑπὸ ΜΘΗ, ΘΗΛ ταῖς ὑπὸ ΘΗΖ, ΘΗΛ ἴσαι
εἰσίν. ἀλλ᾿ αἱ ὑπὸ ΜΘΗ, ΘΗΛ δύο ὀρθαῖς ἴσαι
εἰσίν· καὶ αἱ ὑπὸ ΘΗΖ, ΘΗΛ ἄρα δύο ὀρθαῖς ἴσαι
εἰσίν· ἐπ᾿ εὐθείας ἄρα ἐστὶν ἡ ΖΗ τῇ ΗΛ. καὶ ἐπεὶ
ἡ ΖΚ τῇ ΘΗ ἴση τε καὶ παράλληλός ἐστιν, ἀλλὰ 30
καὶ ἡ ΘΗ τῇ ΜΛ, καὶ ἡ ΚΖ ἄρα τῇ ΜΛ ἴση τε καὶ
παράλληλός ἐστιν· καὶ ἐπιζευγνύουσιν αὐτὰς εὐθεῖαι
αἱ ΚΜ, ΖΛ· καὶ αἱ ΚΜ, ΖΛ ἄρα ἴσαι τε καὶ παράλληλοί εἰσιν· παραλληλόγραμμον ἄρα ἐστὶ τὸ ΚΖΛΜ.
καὶ ἐπεὶ ἴσον ἐστὶ τὸ μὲν ΑΒΔ τρίγωνον τῷ ΖΘ 35
παραλληλογράμμῳ, τὸ δὲ ΔΒΓ τῷ ΗΜ, ὅλον ἄρα
τὸ ΑΒΓΔ εὐθύγραμμον ὅλῳ τῷ ΚΖΛΜ παραλληλογράμμῳ ἐστὶν ἴσον.

Τῷ ἄρα δοθέντι εὐθυγράμμῳ τῷ ΑΒΓΔ ἴσον
παραλληλόγραμμον συνέσταται τὸ ΚΖΛΜ ἐν γωνίᾳ 40
τῇ ὑπὸ ΖΚΜ, ἥ ἐστιν ἴση τῇ δοθείσῃ τῇ Ε· ὅπερ
ἔδει ποιῆσαι.

μϛ΄.

Ἀπὸ τῆς δοθείσης εὐθείας τετράγωνον ἀναγράψαι.

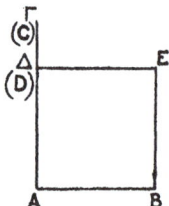

Ε ΣΤΩ ἡ δοθεῖσα εὐθεῖα ἡ ΑΒ· δεῖ δὴ ἀπὸ τῆς
ΑΒ εὐθείας τετράγωνον ἀναγράψαι.
5 Ἤχθω τῇ ΑΒ εὐθείᾳ ἀπὸ τοῦ πρὸς αὐτῇ σημείου
τοῦ Α πρὸς ὀρθὰς ἡ ΑΓ, καὶ κείσθω τῇ ΑΒ ἴση ἡ
ΑΔ· καὶ διὰ μὲν τοῦ Δ σημείου τῇ ΑΒ παράλληλος
ἤχθω ἡ ΔΕ, διὰ δὲ τοῦ Β σημείου τῇ ΑΔ παράλληλος ἤχθω ἡ ΒΕ. Παραλληλόγραμμον ἄρα ἐστὶ τὸ
10 ΑΔΕΒ· ἴση ἄρα ἐστὶν ἡ μὲν ΑΒ τῇ ΔΕ, ἡ δὲ ΑΔ
τῇ ΒΕ. ἀλλὰ ἡ ΑΒ τῇ ΑΔ ἐστιν ἴση· αἱ τέσσαρες
ἄρα αἱ ΒΑ, ΑΔ, ΔΕ, ΕΒ ἴσαι ἀλλήλαις εἰσίν· ἰσόπλευρον ἄρα ἐστὶ τὸ ΑΔΕΒ παραλληλόγραμμον.
λέγω δὴ ὅτι καὶ ὀρθογώνιον. ἐπεὶ γὰρ εἰς παραλ-
15 λήλους τὰς ΑΒ, ΔΕ εὐθεῖα ἐνέπεσεν ἡ ΑΔ, αἱ ἄρα
ὑπὸ ΒΑΔ, ΑΔΕ γωνίαι δύο ὀρθαῖς ἴσαι εἰσίν. ὀρθὴ

δὲ ἡ ὑπὸ ΒΑΔ· ὀρθὴ ἄρα καὶ ἡ ὑπὸ ΑΔΕ. τῶν δὲ
παραλληλογράμμων χωρίων αἱ ἀπεναντίον πλευραί
τε καὶ γωνίαι ἴσαι ἀλλήλαις εἰσίν· ὀρθὴ ἄρα καὶ
ἑκατέρα τῶν ἀπεναντίον τῶν ὑπὸ ΑΒΕ, ΒΕΔ γωνιῶν· 20
ὀρθογώνιον ἄρα ἐστὶ τὸ ΑΔΕΒ. ἐδείχθη δὲ καὶ ἰσόπλευρον.

Τετράγωνον ἄρα ἐστίν· καί ἐστιν ἀπὸ τῆς ΑΒ
εὐθείας ἀναγεγραμμένον· ὅπερ ἔδει ποιῆσαι.

μζ΄.

Ἐν τοῖς ὀρθογωνίοις τριγώνοις τὸ ἀπὸ τῆς τὴν
ὀρθὴν γωνίαν ὑποτεινούσης πλευρᾶς τετράγωνον ἴσον ἐστὶ τοῖς ἀπὸ τῶν τὴν ὀρθὴν
γωνίαν περιεχουσῶν πλευρῶν τετραγώνοις.

ΕΣΤΩ τρίγωνον ὀρθογώνιον τὸ ΑΒΓ ὀρθὴν ἔχον 5
τὴν ὑπὸ ΒΑΓ γωνίαν· λέγω ὅτι τὸ ἀπὸ τῆς ΒΓ
τετράγωνον ἴσον ἐστὶ τοῖς ἀπὸ τῶν ΒΑ, ΑΓ τετραγώνοις.

Ἀναγεγράφθω γὰρ ἀπὸ μὲν τῆς ΒΓ τετράγωνον
τὸ ΒΔΕΓ, ἀπὸ δὲ τῶν ΒΑ, ΑΓ τὰ ΗΒ, ΘΓ, καὶ διὰ 10
τοῦ Α ὁποτέρᾳ τῶν ΒΔ, ΓΕ παράλληλος ἤχθω ἡ ΑΛ·
καὶ ἐπεζεύχθωσαν αἱ ΑΔ, ΖΓ. καὶ ἐπεὶ ὀρθή ἐστιν
ἑκατέρα τῶν ὑπὸ ΒΑΓ, ΒΑΗ γωνιῶν, πρὸς δή τινι
εὐθείᾳ τῇ ΒΑ καὶ τῷ πρὸς αὐτῇ σημείῳ τῷ Α δύο
εὐθεῖαι αἱ ΑΓ, ΑΗ μὴ ἐπὶ τὰ αὐτὰ μέρη κείμεναι 15

τὰς ἐφεξῆς γωνίας δυσὶν ὀρθαῖς ἴσας ποιοῦσιν· ἐπ'
εὐθείας ἄρα ἐστὶν ἡ ΓΑ τῇ ΑΗ. διὰ τὰ αὐτὰ δὴ καὶ

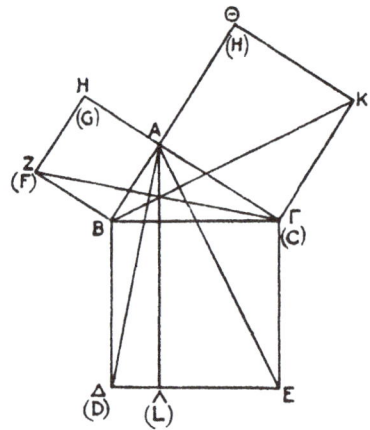

ἡ ΒΑ τῇ ΑΘ ἐστιν ἐπ' εὐθείας. καὶ ἐπεὶ ἴση ἐστὶν
ἡ ὑπὸ ΔΒΓ γωνία τῇ ὑπὸ ΖΒΑ· ὀρθὴ γὰρ ἑκατέρα·
20 κοινὴ προσκείσθω ἡ ὑπὸ ΑΒΓ· ὅλη ἄρα ἡ ὑπὸ ΔΒΑ
ὅλῃ τῇ ὑπὸ ΖΒΓ ἐστιν ἴση. καὶ ἐπεὶ ἴση ἐστὶν ἡ
μὲν ΔΒ τῇ ΒΓ, ἡ δὲ ΖΒ τῇ ΒΑ, δύο δὴ αἱ ΔΒ, ΒΑ
δύο ταῖς ΖΒ, ΒΓ ἴσαι εἰσὶν ἑκατέρα ἑκατέρᾳ· καὶ
γωνία ἡ ὑπὸ ΔΒΑ γωνίᾳ τῇ ὑπὸ ΖΒΓ ἴση· βάσις
25 ἄρα ἡ ΑΔ βάσει τῇ ΖΓ [ἐστιν] ἴση, καὶ τὸ ΑΒΔ
τρίγωνον τῷ ΖΒΓ τριγώνῳ ἐστὶν ἴσον· καί [ἐστι]
τοῦ μὲν ΑΒΔ τριγώνου διπλάσιον τὸ ΒΛ παραλ-
ληλόγραμμον· βάσιν τε γὰρ τὴν αὐτὴν ἔχουσι τὴν
ΒΔ καὶ ἐν ταῖς αὐταῖς εἰσι παραλλήλοις ταῖς ΒΔ,
30 ΑΛ· τοῦ δὲ ΖΒΓ τριγώνου διπλάσιον τὸ ΗΒ τετρά-

γωνον· βάσιν τε γὰρ πάλιν τὴν αὐτὴν ἔχουσι τὴν
ΖΒ καὶ ἐν ταῖς αὐταῖς εἰσι παραλλήλοις ταῖς ΖΒ,
ΗΓ. [τὰ δὲ τῶν ἴσων διπλάσια ἴσα ἀλλήλοις ἐστίν·]
ἴσον ἄρα ἐστὶ καὶ τὸ ΒΛ παραλληλόγραμμον τῷ ΗΒ
τετραγώνῳ. ὁμοίως δὴ ἐπιζευγνυμένων τῶν ΑΕ, ΒΚ 35
δειχθήσεται καὶ τὸ ΓΛ παραλληλόγραμμον ἴσον τῷ
ΘΓ τετραγώνῳ· ὅλον ἄρα τὸ ΒΔΕΓ τετράγωνον δυσὶ
τοῖς ΗΒ, ΘΓ τετραγώνοις ἴσον ἐστίν. καί ἐστι τὸ
μὲν ΒΔΕΓ τετράγωνον ἀπὸ τῆς ΒΓ ἀναγραφέν, τὰ
δὲ ΗΒ, ΘΓ ἀπὸ τῶν ΒΑ, ΑΓ. τὸ ἄρα ἀπὸ τῆς ΒΓ 40
πλευρᾶς τετράγωνον ἴσον ἐστὶ τοῖς ἀπὸ τῶν ΒΑ,
ΑΓ πλευρῶν τετραγώνοις.

Ἐν ἄρα τοῖς ὀρθογωνίοις τριγώνοις τὸ ἀπὸ τῆς
τὴν ὀρθὴν γωνίαν ὑποτεινούσης πλευρᾶς τετράγωνον
ἴσον ἐστὶ τοῖς ἀπὸ τῶν τὴν ὀρθὴν [γωνίαν] περιεχου- 45
σῶν πλευρῶν τετραγώνοις· ὅπερ ἔδει δεῖξαι.

μη΄.

Ἐὰν τριγώνου τὸ ἀπὸ μιᾶς τῶν πλευρῶν
τετράγωνον ἴσον ᾖ τοῖς ἀπὸ τῶν λοιπῶν τοῦ
τριγώνου δύο πλευρῶν τετραγώνοις, ἡ περιεχομένη γωνία ὑπὸ τῶν λοιπῶν τοῦ τριγώνου
5   δύο πλευρῶν ὀρθή ἐστιν.

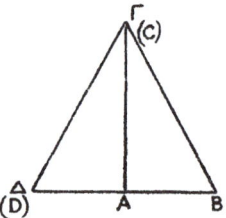

ΤΡΙΓΩΝΟΥ γὰρ τοῦ ΑΒΓ τὸ ἀπὸ μιᾶς τῆς ΒΓ
πλευρᾶς τετράγωνον ἴσον ἔστω τοῖς ἀπὸ τῶν
ΒΑ, ΑΓ πλευρῶν τετραγώνοις· λέγω ὅτι ὀρθή ἐστιν
ἡ ὑπὸ ΒΑΓ γωνία.
10   Ἤχθω γὰρ ἀπὸ τοῦ Α σημείου τῇ ΑΓ εὐθείᾳ πρὸς
ὀρθὰς ἡ ΑΔ καὶ κείσθω τῇ ΒΑ ἴση ἡ ΑΔ, καὶ ἐπεζεύχθω ἡ ΔΓ. ἐπεὶ ἴση ἐστὶν ἡ ΔΑ τῇ ΑΒ, ἴσον
ἐστὶ καὶ τὸ ἀπὸ τῆς ΔΑ τετράγωνον τῷ ἀπὸ τῆς
ΑΒ τετραγώνῳ. κοινὸν προσκείσθω τὸ ἀπὸ τῆς ΑΓ
15   τετράγωνον· τὰ ἄρα ἀπὸ τῶν ΔΑ, ΑΓ τετράγωνα ἴσα
ἐστὶ τοῖς ἀπὸ τῶν ΒΑ, ΑΓ τετραγώνοις. ἀλλὰ τοῖς
μὲν ἀπὸ τῶν ΔΑ, ΑΓ ἴσον ἐστὶ τὸ ἀπὸ τῆς ΔΓ· ὀρθὴ
γάρ ἐστιν ἡ ὑπὸ ΔΑΓ γωνία· τοῖς δὲ ἀπὸ τῶν ΒΑ,
ΑΓ ἴσον ἐστὶ τὸ ἀπὸ τῆς ΒΓ· ὑπόκειται γάρ· τὸ ἄρα

ἀπὸ τῆς ΔΓ τετράγωνον ἴσον ἐστὶ τῷ ἀπὸ τῆς ΒΓ 20
τετραγώνῳ· ὥστε καὶ πλευρὰ ἡ ΔΓ τῇ ΒΓ ἐστιν
ἴση· καὶ ἐπεὶ ἴση ἐστὶν ἡ ΔΑ τῇ ΑΒ, κοινὴ δὲ ἡ
ΑΓ, δύο δὴ αἱ ΔΑ, ΑΓ δύο ταῖς ΒΑ, ΑΓ ἴσαι εἰσίν·
καὶ βάσις ἡ ΔΓ βάσει τῇ ΒΓ ἴση· γωνία ἄρα ἡ ὑπὸ
ΔΑΓ γωνίᾳ τῇ ὑπὸ ΒΑΓ [ἐστιν] ἴση. ὀρθὴ δὲ ἡ 25
ὑπὸ ΔΑΓ· ὀρθὴ ἄρα καὶ ἡ ὑπὸ ΒΑΓ.

Ἐὰν ἄρα τριγώνου τὸ ἀπὸ μιᾶς τῶν πλευρῶν τετράγωνον ἴσον ᾖ τοῖς ἀπὸ τῶν λοιπῶν τοῦ τριγώνου
δύο πλευρῶν τετραγώνοις, ἡ περιεχομένη γωνία ὑπὸ
τῶν λοιπῶν τοῦ τριγώνου δύο πλευρῶν ὀρθή ἐστιν· 30
ὅπερ ἔδει δεῖξαι.

NOTES

NOTES

THE DEFINITIONS

DEFINITION I.

WITH the relative οὗ we have to understand a demonstrative pronoun to which it refers; οὗ = ἐκεῖνο οὗ, 'that of which.' And with μέρος οὐθέν, 'no part,' we must supply ἐστί, 'there is.' The whole sentence then means 'A point is that of which there is no part' or 'A point is that which has no part.' In other words, a point is indivisible; and other writers have expanded the definition by adding that a point 'has no magnitude.'

The Greek word for 'point,' σημεῖον, really means a 'mark,' something, say a dot, to mark where the point is. But the mark itself is not the point. Any mark or dot that we can make, on a sheet of paper, for example, has some size; you can always divide it, or suppose it to be divided, into (say) two parts, which are therefore both smaller than the original mark or dot. This process of division would never come to an end, and some other method is necessary if from a dot which has size and position we are to pass to a point which has position but no size. We can only do so by *abstraction*, i.e. by mentally disregarding the magnitude of the dot and regarding the point as indivisible and having position, but possessed of no other attributes.

This, in substance, had already been stated, before Euclid's time, by Aristotle. Aristotle in fact says that a point is that which is indivisible in respect of quantity and has position, that we can make no distinction between a point and the *place* where it is, and that, a point being indivisible, no accumulation of points, however far it be carried, can give us anything continuous, such as a line; it is only by motion in space that a point can generate a line.

If you cut a line into two parts, or if two lines cut one another, the cutting in either case is at a *point*, which is *where* the line is cut or the lines cut one another; its position or place is determined by the division or intersection (though of course we have in like manner to think of the lines as mathematical lines, i.e. as having no *breadth* whatever). Similarly any line which has an end (or ends) ends in a point, which is simply *where* it ends.

Before Euclid's time the Greeks generally called a point στιγμή, a *puncture* (from στίζειν, 'to prick'). This indicates another way of marking a point, by (let us say) pricking the paper with the point of a very fine needle. But even this puncture, however much smaller it is than any dot we could make with pen or pencil, gives an indentation of *some* size, and is therefore scarcely nearer to an abstract mathematical point than a dot is.

The objection to Euclid's definition is that it states rather what a point is *not* than what it is; we really have to get our conception of a point by other means, namely by the process of abstraction above indicated, before we are in a position to understand Euclid's definition. This is why other ancient writers preferred other definitions. Plato, we are told, objected to recognising points as a separate class of things at all, and regarded them as a 'geometrical fiction.' He preferred to conceive a point as being merely the 'beginning of a line'; alternatively he spoke of 'indivisible lines.' But, as Aristotle says, even indivisible lines must have extremities: hence an indivisible line (even supposing that there is such a thing) must contain at least two points, and cannot therefore be the same thing as a point. The Pythagoreans connected the definitions of a point and a *monad*, or *unit*, saying that a point is a 'monad having position.' The monad or unit being regarded as indivisible, this means something indivisible which has position, and is therefore equivalent to Aristotle's definition above quoted.

DEFINITION 2.

ἀπλατές, neuter of the adjective ἀπλατής (from α- privative and πλάτος, 'breadth'), meaning 'breadthless' or 'without breadth.' 'A line is breadthless length.'

This definition is quoted by Aristotle, and it may safely be attributed to the Platonic school, if not to Plato himself.

The idea of a line is less of an abstraction than that of a point in so far as a line has one positive

8—2

realisable character, namely *length*. As was remarked
in the school of Apollonius of Perga, 'the great
geometer,' we have a notion of a line when we ask
the length of a road or a wall; when we say it is a
mile, or so many feet, long, we ignore all question
of breadth or thickness. A line is, as Aristotle called
it, 'magnitude extended one way' ($\mu\acute{\epsilon}\gamma\epsilon\theta$os $\grave{\epsilon}\phi$' $\grave{\epsilon}\nu$
$\delta\iota\alpha\sigma\tau\alpha\tau\acute{o}\nu$), or, as we say (if the line is straight),
'magnitude in one dimension (only).'

The idea of breadthlessness may be obtained by
considering a line as that which divides a surface or
that which is the intersection of two surfaces (just
as a point is a division of a line or the intersection of
two lines). Consider (1) the division between the
light and the dark when a shadow is thrown on the
ground; if it had had breadth, it would have been
either a part of the dark surface or a part of the
light, but it is neither and is only the division between the two. Similarly (2) with (say) the edge of
a cube; it is the breadthless intersection of two adjoining faces where they meet. But both the line
which divides the dark shadow from the illuminated
area and the edge of the cube have a *length* which
can actually be measured.

Another way of regarding the line was to conceive
it as 'the flux of a point,' i.e. what is described by
a point which moves in space. A material point such
as we can mark on paper has *some* size; hence by its
motion, if it were moved, it would describe a line
which has some breadth or thickness and is visible

to the eye as the material point itself is; but a mathematical point which has *no* size would (if it could be moved) describe a mathematical line, which has no breadth or thickness or visibility.

γραμμή, *line*, of course includes a *curve* as well as a straight line.

DEFINITION 3.

πέρατα, nom. pl. of πέρας, ' limit ' or ' extremity.' Understand ἐστί: ' (the) limits (or ' extremities') of a line are points.'

This remark is an explanation rather than a definition; it is a property of a point, as defined, that it may be the end of a line. Plato, as we have seen, is said to have preferred to regard a point as the 'beginning of a line' rather than as an independent class of thing, a point being, according to him, merely a 'geometrical fiction,' not a real thing. Aristotle alludes to a definition of a point as 'a limit of a line' (πέρας γραμμῆς), but objects to it on the ground that it defines what is prior by means of what is posterior, a point being in the order of thought prior to, or more fundamental than, a line, while a line is similarly prior to a surface and a surface to a solid. Aristotle contrasts what is prior in the order of thought with what is prior *relatively to us*. Relatively to us, a solid is prior to a surface, a surface to a line, and a line to a point. This is because a solid is nearer to sense than a surface is (it is the solid, as Aristotle says, which most of all 'falls

under sense,' i.e. is apprehended by sense), and
similarly a surface is nearer to sense than a line, and
a line than a point. Consider a solid cube, say a
piece of metal in that shape; it is a thing which we
can see and feel; and a mathematical cube is the
same thing with all but its geometrical attributes
omitted, but it still has three dimensions. Any one
of the faces is a surface, but this is more remote from
sense, since we have to abstract all idea of depth or
thickness and to think of an area with no thickness.
A further step in abstraction is necessary in order
to apprehend a line in what we call an 'edge' of
the cube, one side of the plane face; for we have to
abstract the breadth of the face and leave only its
(breadthless) length. A third step in the process of
abstraction is required in order to apprehend a
mathematical *point* in the corner of the cube which
is one end of the edge which we have taken.

Euclid, in giving this definition, shows that, while
he prefers, on principle, the scientific definition of
a point, he recognises another, the current, view of
it, which, as we have seen, was that of Plato.

DEFINITION 4.

This famous definition is full of difficulty; it is
difficult to translate the Greek, and it is not less
difficult, when we have translated the words, to
assign an intelligible meaning to them. 'A straight
line is any (line) which (ἥτις) lies (κεῖται) ἐξ ἴσου
κ.τ.λ.' The words ἐξ ἴσου (where the adjective

ἴσος has almost the force of a substantive) are generally used by Plato and Aristotle in the sense of 'on a footing of equality'; once Aristotle speaks of 'asking a question ἐξ ἴσου,' which means asking a question impartially, or without bias one way or the other, i.e. without showing any expectation of one answer being given rather than another. These uses suggest that the words must here mean 'evenly placed,' 'without bias,' i.e. without inclining one way or the other. Hence 'evenly' is not a bad translation of ἐξ ἴσου. Next, the dative τοῖς ἐφ' ἑαυτῆς σημείοις seems to be constructed with ἐξ ἴσου rather than with κεῖται, and we have the following as the result: 'a straight line is any (line) which lies evenly with the points on itself.'

Simson substituted for 'the points on itself' the words 'its extreme points,' no doubt in order to make the signification clearer; but Euclid's words are more general, since they take into account, not only the ends of the line (which would not even exist if the line happened to be unlimited or infinite in length), but *all* the points on it. That is, Euclid says that the straight line lies evenly with respect to any two (or any number of) points on it; which may be taken to mean that, whatever portion of a straight line is taken, say between two points on it, it shows with reference to these points no swerve or unevenness towards one side or the other; both sides are alike, in the sense that, if the line were supposed to be turned over by revolution about the

two points as poles, or turned the reverse way, as
between the two points, with or without such
revolution, it would still occupy precisely the same
position (which would not be the case if it had any
bend towards either or any side in any part of its
length).

In modern text-books the idea is explained by
illustrations rather than by a formal definition. If
we consider a stretched string, e.g. a plumb-line, or
a ray of light entering a room through a small hole,
we get a notion of a straight object, and by elimina-
ting all thickness from this straight object we arrive
at the abstract conception of a mathematical straight
line, of which the straight line that we draw is an
imperfect representation.

The only pre-Euclidean definition of a straight
line which has been handed down is that given by
Plato, who called it 'that of which the middle
covers (literally 'stands in front of,' i.e. obscures
the view of) the ends,' relatively, that is, to an eye
placed at either end and looking along the line.
This implies an appeal to sight, which would verify
(if the straight line could be *seen* end-on) that there
is no swerve in any part of the straight line in any
direction. Plato's notion would therefore appear to
be in effect the same as that of Euclid according to
the above interpretation. It seems, in fact, probable
that Euclid got his notion from Plato's definition,
but, as he saw that a purely geometrical definition
should be independent of any sense-perception

(which belongs to physics rather than geometry), he tried his best to express the same thing in general terms without any appeal to sight.

Other ancient definitions are given by Heron of Alexandria and Proclus, e.g. (1) 'a line stretched to the utmost'; (2) 'a line such that all its parts fit on all (other parts) in all ways'; (3) 'that line which, when its extremities remain fixed, itself remains fixed,' i.e. the line which does not change its position when it is turned about its extremities as poles (this is equivalent to definitions given in modern times by Leibniz, Saccheri and Gauss).

The definition of a straight line as 'the shortest distance between two points' is also ancient; the fact is stated, not as a definition, but as an 'assumption' (λαμβανόμενον), by Archimedes. This definition again has been used, in modern times, by Legendre.

These alternative definitions, so-called, as well as Archimedes's assumption, are really not definitions but rather statements of properties of a straight line consequent on a straight line being, as Euclid says, 'any line which lies evenly with the points on itself.'

DEFINITION 5.

ἐπιφάνεια, 'surface,' means literally that feature of a body (the outside) which is *apparent* to the eye (ἐπιφανής, from ἐπί and the root of φαίνεσθαι, 'to appear').

ὅ = ἐκεῖνο ὅ, 'that which.'

We are told that the Pythagoreans called a surface χροιά, which seems to have meant *skin* as well as *colour*. There is this much of appropriateness in the word that it is the colour of the surface which enables it to be seen.

The definition of a surface corresponds to that of a line. As a line has length but no breadth or depth, a surface has length and breadth but no depth. Aristotle calls it 'magnitude extended, or continuous, *two ways*' (ἐπὶ δύο or διχῇ); in the same way, if a surface is a plane surface, we say that it is in two dimensions. As a line is terminated or divided at a point, and a surface in a line, so a solid is terminated or divided by a surface.

We get an idea of a surface by considering, say, the surface of a piece of water which is the boundary between the mass of water below and the air above, but has no thickness and is neither water nor air (if it had thickness it would have to be one or other, water or air); and similarly by looking at a shadow on the ground, since a shadow has no depth (it does not penetrate the earth).

As a point by its motion in space generates a line, so a line by its motion in space (otherwise than along its own length) generates a surface.

DEFINITION 6.

The remarks on Def. 3 apply *mutatis mutandis* to this definition. Euclid does not *define* a line as a limit, extremity, or division, of a surface; this, as

Aristotle says, would be unscientific; but he recognises this aspect of a line by adding the remark in Def. 6 as an explanation, not as a definition.

DEFINITION 7.

ἐπίπεδος, 'plane' (adjective, from ἐπί and the root πεδ; cf. πούς, pes, a foot). The etymology of the word shows that the idea is that of something you can *stand on*, i.e. flat. ἐπίπεδον, the neuter, by itself, means a 'plane.'

This definition follows exactly, *mutatis mutandis*, the definition of a straight line, ἐπίπεδος ἐπιφάνεια, 'plane surface,' taking the place of εὐθεῖα γραμμή, 'straight line,' and ταῖς...εὐθείαις, 'the straight lines,' replacing τοῖς...σημείοις, ' the points.' Hence we must translate thus: 'A plane surface is any (surface) which lies evenly with the straight lines on itself.' The explanation, too, of the definition must correspond to that of Def. 4. Proclus tells us that 'a plane surface is a surface the middle of which covers the ends.' This is an adaptation of Plato's definition of a straight line. What is meant is that, if we could *see* a straight line, and if we were to place any two straight lines in the plane in such a position that one line exactly covered the other as we looked at them (this assumes that the straight lines *can* be so placed), the plane would be seen as a mere line coincident with the nearest of the two lines; no part of the plane would bulge on either side of the straight line seen by us. Euclid, then, seems to have tried to express the same fact in general terms, without any appeal to sight.

Simson substituted in his Euclid a different definition to the effect that 'A plane surface is that in which, any two points being taken, the straight line joining them lies wholly in the surface,' a definition which is equivalent to others found in Heron and Theon of Smyrna. Archimedes states, as an *assumption*, that, of all surfaces which have the same extremities, the plane surface is the least (in area): this corresponds to his remark that, of all lines connecting two points, the straight line is the least.

DEFINITION 8.

The words of this definition should, for the purpose of the translation, be taken in the following order: ἐπίπεδος γωνία ἐστὶν ἡ κλίσις πρὸς ἀλλήλας δύο γραμμῶν ἐν ἐπιπέδῳ ἁπτομένων ἀλλήλων καὶ μὴ ἐπ' εὐθείας κειμένων. κλίσις (from κλίνειν) means 'inclination' or 'leaning.' ἁπτομένων (pres. part. of ἅπτεσθαι, 'to meet,' not as a rule 'to *touch*,' which is ἐφάπτεσθαι) and κειμένων (pres. part. of κεῖσθαι, 'to lie' or 'to be placed') both agree with γραμμῶν, and ἀλλήλων is the objective genitive after ἁπτομένων, 'meeting *one another*.' πρὸς ἀλλήλας goes with κλίσις δύο γραμμῶν, the 'inclination of two lines *to one another*.' We translate therefore: 'A plane angle is the inclination to one another of two lines in a plane which meet one another and do not lie in a straight line.' The words τῶν γραμμῶν before κλίσις are unnecessary and should be ignored;

they were no doubt added in view of the separation of
κλίσις from δύο γραμμῶν by as many as nine words.

The definition purports to cover an angle included by a curve and a straight line, or by two
curves, as well as an angle formed by two straight
lines. This is evident from the fact that the *rectilineal* angle is separately described in the next
definition. This being so, the words 'which do not
lie in a straight line' are strange, since, if the lines
containing the angle are curves, it would be more
appropriate to say 'which are not *continuous* with
one another.' It looks as though Euclid really
intended to define a rectilineal angle in this definition and then, on second thoughts, as a con-
cession to the then common recognition of curvi-
lineal angles, altered 'straight lines' into 'lines' and
separated the definition into two.

The evidence suggests that Euclid's definition of
an angle as an 'inclination' was a new departure.
The word κλίσις does not occur in Aristotle, and we
gather from him that the idea commonly associated
with an angle in his time was rather *deflection* or
breaking of lines (κλάσις), i.e. a *bend* made by a
broken line. Apollonius defined an angle as a 'con-
tracting of a surface or a solid at one point under a
broken line or surface (respectively)'; the aspect of
an angle here emphasised is *convergence* to a point.
The converse aspect of an angle as the *divergence* of
lines from a point was also recognised; for some
writers, including Plutarch, said that 'the first dis-

tance under the point' (τὸ πρῶτον διάστημα ὑπὸ τὸ σημεῖον) is the angle. To realise this, take two points on the lines respectively at equal distances from the point of intersection and very near to it, and consider 'the distance under the point' as being the distance between the two points so taken; we have then a sort of measure of the *rate of divergence* and therefore of the size of the angle.

DEFINITION 9.

The straight lines which form an angle are said to 'contain' or 'enclose' (περιέχειν) the angle: αἱ περιέχουσαι τὴν γωνίαν γραμμαί are therefore 'the straight lines containing the angle.' εὐθύγραμμος (from εὐθύς, 'straight,' and γραμμή, 'line') is *rectilineal*, i.e. formed by straight lines.

DEFINITION 10.

σταθεῖσα, 'set up,' aorist passive participle of ἱστάναι. ἐφεξῆς, an adverb meaning 'successively,' 'next in order.' As Aristotle says, 'that is successive which is after the beginning and has nothing of the same kind between it and that to which it succeeds.' The word is technically used, as here, in the expression αἱ ἐφεξῆς γωνίαι, to mean 'the adjacent angles.' ὀρθὴ ἑκατέρα τῶν ἴσων γωνιῶν ἐστί, 'each of the equal angles is right,' i.e. is a right angle.

ἡ ἐφεστηκυῖα εὐθεῖα, 'the straight line standing (on the other),' ἐφεστηκυῖα being the fem. of the perf. participle active (used in the intransitive sense) from ἐφιστάναι.

κάθετος, *perpendicular*, means literally 'let fall' (verbal adjective from καθιέναι); the full expression is ἡ κάθετος εὐθεῖα γραμμή, 'the perpendicular straight line.' The notion is that of a straight line let fall on the surface of the earth, a *plumb-line*. We learn from Proclus that in ancient times the perpendicular was described as drawn *gnomon-wise* (κατὰ γνώμονα), because the gnomon (an upright stick, and hence the perpendicular needle of a sun-dial) was set up at right angles to the horizon.

ἐφ' ἦν = ἐπὶ τὴν γραμμὴν ἐφ' ἦν, the first ἐπὶ being constructed with κάθετος, 'perpendicular' (as we say) '*to* the straight line on which it stands.'

The full translation then is: 'When a straight line set up on a straight line makes the adjacent angles equal to one another, each of the equal angles is right, and the straight line standing on the other is called a *perpendicular* to that on which it stands.'

DEFINITION II.

ἀμβλύς is opposed to ὀξύς, 'sharp,' and means 'blunt,' i.e. with the edge or point taken off or blunted; hence, when used of an angle, *obtuse*. We may translate 'an obtuse angle is *an* angle greater than a right (angle),' although the Greek idiom prefers the definite article in such cases, '*the* (angle which is) greater than a right angle.' ἡ μείζων ὀρθῆς is of course really the subject: '*the* angle (or *that* angle) which is greater than a right angle is obtuse' would be the literal translation.

DEFINITION 12.

ὀξύς, 'sharp,' or, when used of an angle, 'acute.' Aristotle discusses the question whether the right angle is *prior* to the acute or *vice versa*. He concludes that the right angle is prior in *notion* and in being *defined*; the definition of a right angle is independent, whereas an acute angle is defined by means of a right angle. Only as *matter* could the acute angle be called prior, i.e. in the sense that a right angle can be made up as the sum of a number of acute angles (two or more), in which case the acute angles are parts of the right angle, and in that sense are the material of which it is made up.

DEFINITION 13.

ὅρος, 'boundary.' The same word is used in Greek for 'definition'; 'that which *defines*' would in English, also, be true of a boundary. ὅ = ἐκεῖνο ὅ, 'that which.'

πέρας, 'limit,' or 'extremity,' as usual. With Aristotle ὅρος and πέρας are synonymous.

DEFINITION 14.

σχῆμα, 'figure'; περιεχόμενον, 'contained' or 'enclosed,' as usual.

Before Euclid's time ideas of 'figure' were more vague. Thus Plato says that *roundness* or *the round* (circular) is a figure, and that *the straight* is so too. Aristotle too classes *angle, straight, circular* as species of figure. Plato and Aristotle seem however to be here speaking of 'figure' in the sense of *shape* rather than of 'figure' in our sense. Plato comes nearer to Euclid's idea of figure when he says that 'figure is an extremity of a solid,' 'that in which the solid terminates.' Aristotle too elsewhere uses language not unlike Euclid's, distinguishing plane figures of two kinds, those *contained* by straight and circular lines respectively. The definition of ' figure ' is however probably Euclid's own. It is clear from his use of ὅρος, 'boundary,' in the definition that he would not regard a straight line as being a figure, and probably not an angle either.

Posidonius defined 'figure' in the very short phrase πέρας συγκλεῖον, 'confining limit' or 'extremity.'

Since Euclid says that a figure is '*that which is contained* by any boundary or boundaries,' he includes in the idea not merely the boundary or contour but what is inside it, the content, as well.

DEFINITION 15.

κύκλος, ' circle '; σχῆμα ἐπίπεδον, 'a plane figure,' i.e. a figure lying in a plane.

γραμμῆς, 'line,' here used in the sense of a curve.

'A circle is a plane figure contained by one line.' The MSS. add, in further explanation of the one 'line' which encloses the circle, the words ἣ καλεῖται περιφέρεια, 'which (line) is called *circumference*.' But it is probable that these words were interpolated by some one who had in view the occurrence of the word περιφέρεια without any explanation in Defs.

17 and 18 following. The word περιφέρεια however, in the sense of *contour*, without any special mathematical signification, was well understood before Euclid's time, and he was therefore entitled to use it in its ordinary meaning without defining it mathematically. The meaning is of course that of something *carried round* (περί and φέρω); cf. the adjective περιφερής, 'round' or 'circular.'

πρὸς ἣν, 'to which' (sc. γραμμήν), is constructed with προσπίπτουσαι (προσπίπτω), 'falling on which,' i.e. 'drawn to which.'

ἀφ' ἑνὸς σημείου τῶν ἐντὸς τοῦ σχήματος κειμένων, 'from one point of (i.e. among) the (points) lying within the figure.' A literal translation of the relative clause πρὸς ἣν...ἴσαι ἀλλήλαις εἰσίν would be intolerable in English ('on which all the straight lines which fall from one point among those lying within the figure are equal to one another'), and we have to say *'which is such* that all the straight lines falling upon *it* from one point among those lying within the figure are equal to one another.' The words πρὸς τὴν τοῦ κύκλου περιφέρειαν after εὐθεῖαι are obviously interpolated and should be ignored in translating.

Passing from the definition of *figure* in general to definitions of particular figures, Euclid naturally gives first the definition of the figure which has only *one* boundary, being bounded by one line, and then proceeds to take in order the figures which have more than one boundary (two, three, four, etc.).

The definition of a circle contains nothing that was new in substance. The same thing had been said before in terms less formal but more terse. Thus Plato says '*Round* (i.e. circular) is, I take it, that the extremes of which are every way equally distant from the middle'; and Aristotle uses the expression 'the plane equal (i.e. extending equally all ways) from the middle' for a circle, while he speaks elsewhere of 'the circular (literally 'circumferential-lined,' περιφερόγραμμον) plane figure bounded by one line.'

Observe that with Euclid and the Greek mathematicians generally the circle is not merely what we see drawn, the circumference, but the whole of the plane figure included by it. Thus where we might speak of a straight line meeting a circle the Greek would be careful to say that the straight line meets the circumference of the circle. There are however a few exceptions even in Euclid; cf. III, 10 'A circle does not cut a circle in more points than two.'

Observe also that the definition says nothing on the question whether any figure answering to the definition can exist or not. This is not stated until Postulate 3, where the possibility of drawing a circle, with a given centre and 'distance,' is assumed. It is true to say of definitions in general that they state nothing as to the existence or non-existence of the thing defined. The definition explains the meaning of a term, and it is only necessary that it should be understood. The fact that the thing defined exists has to be proved (or postulated) later, and in geometry is generally proved by actual construction. Thus that a right angle exists does not follow from Def. 10 but only from Prop. 11 of Book 1, where it is actually drawn. Similarly with a square, which is defined in Def. 22, but is not shown to exist till it is actually drawn in 1, 46.

DEFINITION 16.

κέντρον, 'centre.' 'And the point' (namely the point within the circle which is such that all the straight lines drawn from it to the circumference are equal) 'is called (the) centre of the circle.' The word κέντρον was regularly used even before Euclid's time; it means literally that which makes a puncture (κεντεῖν, 'to prick'), e.g. the stationary leg of a pair of compasses when used to draw a circle; hence it comes to be used for the place (the centre) where the puncture is made.

DEFINITION 17.

διάμετρος (ἡ), *diameter*, literally 'that which measures through' (διά and μετρεῖν), was the regular word, in Euclid and in the classical Greek writers, for the 'diameter' of a square or a parallelogram as well as for the diameter of a circle; διαγώνιος, 'diagonal,' i.e. passing through from angle to angle (διά and γωνία), was a later term.

εὐθεῖά τις, 'any straight line'; ἠγμένη, 'drawn' (fem. of the perf. participle passive from ἄγειν, a word regularly used for 'drawing' lines in geometrical constructions); περατουμένη (περατόω), 'limited' or 'terminated' (cf. πέρας, 'limit' or 'extremity').

ἐφ' ἑκάτερα τὰ μέρη, 'in both directions,' literally 'towards both the parts,' where 'parts' must be used in the sense of 'regions' (cf. Thuc. II, 96).

ἥτις, best translated 'and any such (straight line),' i.e. any diameter.

δίχα τέμνει, 'bisects,' literally 'cuts in two'; δίχα τέμνειν, like δίχα διαιρεῖν, regularly implies cutting into two *equal* parts.

We thus arrive at the translation 'A diameter of the circle is any straight line drawn through the centre and terminated in both directions by the circumference of the circle, and any such straight line also bisects the circle.'

Proclus relates that the first to prove that a circle is bisected by its diameter was Thales. We are not told how he proved it. Presumably the proof would be based on the *symmetry* of the circle with respect to any diameter, however this was brought out, e.g. by turning one of the portions into which the diameter divides the circle over, *applying* it to the other portion, and proving that the two portions coincide exactly when so applied.

The last words 'and any such (straight line) also bisects the circle' are omitted by Simson and the editors who followed him as not belonging to the definition but stating a property of the diameter as defined. All the same they are necessary where they occur; for, without this explanation, Euclid would not have been justified in describing as a *semi*-circle (in Def. 18) a portion of a circle bounded by a diameter and the circumference cut off by it.

DEFINITION 18.

ἡμικύκλιον, 'semicircle' (ἡμι= *semi* and κύκλος). The construction is (τὸ) σχῆμα τὸ περιεχόμενον ὑπὸ κ.τ.λ., 'the figure contained by the diameter and....'

ἀπολαμβανομένης, 'cut off' (pres. participle pass. of ἀπολαμβάνειν), agreeing with περιφερείας. ὑπ' αὐτῆς, 'by it,' i.e. by the diameter.

Observe that περιφέρεια is used here, as often, for a *part* of the circumference, i.e. what we call an 'arc' of a circle.

The full translation is 'A semicircle is the figure contained by the diameter and the circumference cut off by it.'

The last words, κέντρον δὲ τοῦ ἡμικυκλίου τὸ αὐτό, ὃ καὶ τοῦ κύκλου ἐστίν, 'And the centre of the semicircle is the same as that of the circle'—literally 'the centre of the semicircle is the same (point) which is also (the centre) of the circle'—come from Proclus's commentary and not from the mss. Proclus remarks, rather absurdly, that the semicircle is the only plane figure that has its centre on its perimeter (!). The fact is that, as Scarburgh said in *The English Euclide*, a semicircle as such has no centre, properly speaking; the 'centre' is not the centre of the semicircle but only of the circle of which it forms part. The words are probably better left out altogether.

In defining particular figures Euclid (as we said) takes them in an order corresponding to the number of their boundaries. He begins with the figure (i.e. the circle) which has *one* boundary; he proceeds then to the figure which has *two*, namely the semicircle (which is bounded by its circumference and its diameter); and after that he takes the figures which are bounded by three or more lines, in order. All the figures defined in Defs. 19–22 are bounded by *straight* lines and are therefore included in the class of *rectilineal* figures.

εὐθύγραμμος (from εὐθύς and γραμμή), *rectilineal*, meaning, as Euclid immediately explains, 'contained by straight lines.'

τρίπλευρον, 'three-sided' or 'trilateral' (πλευρά = 'side'); τετράπλευρον, 'four-sided' or 'quadrilateral'; πολύπλευρον, 'many-sided' or 'multilateral.'

With ὑπὸ τριῶν, ὑπὸ τεσσάρων, ὑπὸ πλειόνων ἢ τεσσάρων understand εὐθειῶν: πλειόνων ἢ τεσσάρων, 'more than four.'

The latter part of this definition distinguishing 'three-sided,' 'four-sided' and 'many-sided' figures is probably due to Euclid himself, since the words τρίπλευρον, τετράπλευρον and πολύπλευρον do not appear in Plato or in the genuine Aristotelian writings. By his use of τετράπλευρον, *quadrilateral*, Euclid no doubt wished, once for all, to put an end to any ambiguity in the use of the word τετράγωνον, literally a 'four-angled (figure),' which had some-times been used in the general sense of any four-sided figure, and to get it formally restricted to the *square*, for which it is the regular term.

DEFINITION 20.

The regular word for 'triangle' is τρίγωνον, literally 'three-angled (figure),' and this definition distinguishes triangles as being of three kinds according to the relations between the lengths of the sides, while the next definition classifies them with reference to their angles.

ἰσόπλευρον, 'equal-sided' or 'equilateral.'

In translating τὸ τὰς τρεῖς ἴσας ἔχον πλευράς we take the words in the order τὸ ἔχον τὰς τρεῖς πλευρὰς ἴσας, 'that (triangle) which has its (literally 'the') three sides equal.' Similarly in the definition of the second kind of triangle (the isosceles) we take the words in the order τὸ ἔχον τὰς δύο πλευρὰς μόνας ἴσας, 'that (triangle) which has two of its sides (literally 'the two sides') alone equal'; and in the definition of the third kind of triangle we translate the words in the order τὸ ἔχον τὰς τρεῖς πλευρὰς ἀνίσους.

ἰσοσκελής, *isosceles*, meaning 'with equal *legs*' (σκέλος). The word is used by both Plato and Aristotle.

σκαληνός, *scalene*, also used by Aristotle of a triangle with no two sides equal. Proclus seems to connect the word with σκάζειν, 'to limp'; others make it akin to σκολιός, 'crooked,' 'aslant.' Apollonius uses the word *scalene* of an *oblique* circular cone. Plato in one place applies the term 'scalene' to an *odd* number in contrast to 'isosceles' used of an even number (*Euthyphro*, 12 D).

DEFINITION 21.

ὀρθογώνιος (from ὀρθός and γωνία), 'right-angled'; ἀμβλυγώνιος, 'obtuse-angled,' and ὀξυγώνιος, 'acute-angled,' are similarly formed.

The descriptions of these triangles respectively require careful translation. τὸ ἔχον ὀρθὴν γωνίαν, 'that (sc. triangle) which has *a right angle*'; τὸ ἔχον ἀμβλεῖαν γωνίαν, 'that which has *an obtuse angle*'; τὸ τὰς τρεῖς ὀξείας ἔχον γωνίας (which we must take in the order τὸ ἔχον τὰς τρεῖς γωνίας ὀξείας), 'that which has its three angles acute.' The difference in form between the first two expressions and the third is due to the fact that, while an acute-angled triangle has all its angles acute, the other triangles can only have *one* angle right or *one* angle obtuse, as the case may be, and the other two angles must in either case be acute. This follows from the fact (proved in I, 32) that the sum of the three angles of any triangle is equal to two right angles.

DEFINITION 22.

Quadrilaterals are now classified with reference to the attributes 'equilateral' and 'right-angled' or the absence of them respectively. The alternatives are clearly that a quadrilateral may be

(1) both equilateral and right-angled: this is a *square*;

(2) right-angled but not equilateral: Euclid says that this quadrilateral is an oblong (ἑτερόμηκες);

(3) equilateral but not right-angled: a *rhombus*; or

(4) neither equilateral nor right-angled.

Dealing with the last class, Euclid distinguishes, from the rest, the figures which, though they have not *all* their sides equal, have the opposite sides equal respectively, and the opposite angles equal respectively, in pairs. These figures he calls by the name *rhomboid*. All the rest of the class he calls by the general name *trapezium* (τραπέζιον, diminutive from τράπεζα, a table, i.e. 'a little table').

It is clear that 'right-angled' in this definition means 'having all its angles right,' because, although for a square, which has all its sides equal, it is sufficient to say that *one* of its angles is right, this is not the case with the oblong which is not equilateral. As a matter of fact, the quadrilateral figure (a square) which has all its sides equal and one angle right has *all* its angles right; but the quadrilateral which has its sides unequal and one angle right has not necessarily any more of its angles right, whereas what Euclid calls an *oblong* is really what we call a rectangle, a parallelogram with all four angles right.

τετράγωνον, literally 'four-angle,' was already a *square* with the Pythagoreans, and it is so most commonly in Aristotle; but in a few passages Aristotle evidently uses the word with the wider signification of 'quadrilateral.' By introducing the new word τετράπλευρον for the quadrilateral in general Euclid wished no doubt to confirm the conventional restriction of τετράγωνον to a *square*.

ἑτερόμηκες means literally a figure with sides *of different lengths*, but it was conventionally appropriated even by the Pythagoreans to the *oblong*, in other words a *rectangle*.

ῥόμβος, *rhombus*, is apparently derived from ῥέμβειν, 'to turn round and round'; it meant among other things a 'spinning-top.'

ῥομβοειδές, *rhomboid*, literally 'rhombus-shaped.'

ἀπεναντίον, 'opposite' (cf. ἐναντίος), an adverb used as an adjective qualifying πλευράς and γωνίας respectively: 'that which has its opposite sides and angles respectively equal to one another.' The rhomboid is really a parallelogram. Observe here too that it is not yet proved that any such quadrilateral exists; that it exists is first proved in I, 34.

Euclid does not anywhere define a parallelogram. The definition in any case could not have preceded the definition of parallel straight lines (Def. 23); but, when we come to parallelograms in Book I,

Euclid introduces them as 'parallelogrammic areas or, more exactly, 'parallelogram areas' (παραλληλό-γραμμα χωρία), evidently regarding the expression as self-explanatory (= areas contained by parallel lines); after which, in the same proposition (I, 34), he shortens the expression into 'parallelogram,' a word which is, in reality, still an adjective with χωρίον understood, although we make a substantive of it.

Though Euclid, in the above definition, classes as *trapezia* all quadrilaterals which are not squares, oblongs, rhombi or rhomboids, the term was generally restricted by later writers to quadrilaterals having two sides (only) parallel. It is used in this sense by Archimedes and Heron, and apparently by Euclid himself in his book *On divisions (of figures)*.

Euclid makes no use in the *Elements* of the terms *oblong, rhombus* and *rhomboid*. The explanation of his inclusion of definitions of these figures is probably that they were taken from earlier text-books.

DEFINITION 23.

παράλληλος, 'parallel,' means literally 'alongside one another' (παρά and ἀλλήλας).

ἐκβαλλόμεναι, 'being *produced*.' ἐκβάλλειν ('throw out') is the regular expression in Greek geometry for producing a straight line.

εἰς ἄπειρον. ἄπειρος (a- privative) is connected with πέρας, 'limit' or 'end,' and means 'without limit' or 'infinite.' εἰς ἄπειρον is adverbial, meaning 'without limit' or, as mathematicians commonly say, 'indefinitely.' Simson translated 'ever so far,' which of course is not far enough.

ἐφ' ἑκάτερα τὰ μέρη, 'in both directions,' as in Def. 17.

ἐπὶ μηδέτερα: understand τὰ μέρη, 'in neither direction.'

'Parallel straight lines are any (straight lines) which, being in the same plane and being produced without limit in both directions, do not meet one another in either direction.'

Another view of parallel straight lines held in antiquity was that they are such that, however far they are produced in both directions, they always remain the same distance apart. The distance between them is of course the *shortest* distance, which is measured by a straight line perpendicular to both the straight lines. Proclus quotes a definition given by Posidonius to this effect: 'Parallel straight lines are those which, (being) in one plane, neither converge nor diverge, but have all the perpendiculars equal which are drawn from the points of the one (straight line) to the other.' This definition is however (as Saccheri pointed out) unsatisfactory because, before such a definition can be used, it has to be *proved* that the geometrical locus of points equidistant from a given straight line is also a straight line.

It is clear that with Aristotle the general notion of parallels was that of straight lines *which do not meet*, as in Euclid.

THE POSTULATES

WE come now to five assumptions known as Postulates. The Greek term is αἴτημα, a *demand* or *postulate*, formed from αἰτεῖν, 'to ask' or 'demand.' Euclid does not here use the substantive but a part of the verb, namely ᾐτήσθω, 3rd pers. perfect imperative passive. This use of the *perfect* imperative passive is very characteristic of Greek geometry, and it is very expressive. Thus ᾐτήσθω means literally 'Let it *have been* demanded,' not merely 'Let it be demanded.' Similarly, in constructions, we have γεγράφθω, 'let it have been described' or 'drawn,' and ἤχθω (from ἄγειν) 'let it have been drawn'; that is to say, the one word means more than 'let it be described (or drawn)'; it is equivalent to 'suppose it described (or drawn).' This elegant and concise mode of expression is not possible in English which has not the richness in inflexions that the Greek has; and we have in practice no alternative but to translate by the present imperative, 'let it be demanded,' and the like. In the present case, in order to keep the technical term *postulate*, it is convenient to say 'Let the following be postulated.'

The nature of a Postulate is well brought out by Aristotle, who discusses at considerable length the terms *definition, hypothesis, postulate* and *axiom* in relation to one another. In demonstrative sciences

such as geometry it is necessary to begin from certain
assumptions which are not capable of demonstration: otherwise there would be an infinite series
of demonstrations. One class of indemonstrable assumptions is that of 'axioms' (in Euclid 'Common
Notions'), for which Aristotle also uses the alternative terms 'common (things)' (τὰ κοινά) or
'common opinions' (κοιναὶ δόξαι). Axioms are
general principles, common to all sciences, and self-
evident though incapable of proof, e.g. that 'one
of two contradictories must be true' or that 'when
equals are subtracted from equals, the remainders
are equal.' Any person capable of reasoning at all
must accept such assumptions as these without
proof, and every one must grasp and hold them
firmly if he is to learn anything at all. A *postulate* is
also a necessary assumption, but it differs from an
axiom in that it is not self-evident, or a *necessary
truth*; the learner need not be convinced of it by
the reason that is within him; it is a thing assumed
by the learner on the authority of the teacher only,
and the learner himself may have no opinion of his
own on the subject, or the postulate may actually
be rather contrary than otherwise to his opinion if
he has one. This description seems to fit the Postulates of Euclid accurately enough; they are *demands*
made without any assent on the part of the learner,
but they are necessary as a basis if a science is to get
under way at all.

The first three Postulates are generally called the

Postulates of construction. But they are more than that. Aristotle had remarked that, before we can have a demonstrative science dealing with points and lines, we must be satisfied that points and lines *exist*. The existence, then, of straight lines and circles is postulated by Euclid in the form of a categorical assumption that such lines can be *constructed*, the actual construction of a thing (if it can be effected) being a sufficient proof that it exists. There is a further point. No straight line or circle that we can draw with the imperfect instruments at our disposal is a perfect (mathematical) straight line or circle satisfying the respective definitions. Euclid was aware of this, and yet he deliberately assumes that mathematical straight lines and circles *can* be drawn. By this he implies that the imperfection of our instruments and the inaccuracy of the figure actually drawn *do not matter*. We are entitled to suppose that the true straight line or circle is there, although we have not been able to draw it accurately; the validity of the proof is not affected by the imperfection of the figure, because the proof is not concerned with the particular imperfect straight lines and circles which we have drawn but with the real straight lines and circles of which those that we draw are only illustrations. As Aristotle says, a geometer may quite legitimately call a line a foot long when it is not, or straight when it is not straight; he is not on that account using hypotheses which are false.

With these general observations, we pass to the individual Postulates.

POSTULATE I.

ἀπὸ παντὸς σημείου ἐπὶ πᾶν σημεῖον, literally 'from *every* point to *every* point.' In such cases as this the Greeks speak of *every* where we say *any*. The meaning of ' every ' here is of course ' any whatever' or 'any we please.' Cf. the enunciation of I, 18, 'In *every* triangle the greater side subtends the greater angle.'

ἀγαγεῖν, aorist infinitive of ἄγειν, 'to draw.'

The full translation is 'Let the following be postulated: to draw a straight line from any point to any point' (i.e. that we can draw, etc.).

The statement that we can draw a straight line from any one point to another implies that such a straight line exists; but the Postulate has yet another implication, namely that the operation is *one*, and that the straight line joining the two points is *unique*: we cannot draw two (or more) different straight lines joining two points, and this again implies that *two straight lines cannot enclose a space*. Consequently the axiom to the latter effect which was interpolated among the genuine Euclidean axioms is unnecessary. That Euclid had no axiom to this effect is clear from I, 4, in the proof of which he tacitly assumes that two straight lines cannot enclose a space without referring to any such axiom.

POSTULATE 2.

πεπερασμένην, 'limited' or 'terminated,' that is to say, a straight line which has *ends*. It is possible also to translate πεπερασμένην by 'finite,' just as ἄπειρος, 'unlimited,' may also be translated by 'infinite.'

κατὰ τὸ συνεχές, literally 'according to the continuous' or 'continuity'; i.e. 'continuously.'

ἐπ' εὐθείας, literally 'on a straight line,' i.e. 'in a straight line'; ἐκβαλεῖν, aor. infin. of ἐκβάλλειν, 'to produce,' as usual.

'To produce a terminated straight line continuously in a straight line.'

The Postulate implies that the produced portion of the straight line is *unique,* or, in other words, that *two straight lines cannot have a segment common to both.*

Attempts were made in ancient as well as in modern times to prove this latter proposition. Proclus and Simplicius attempted it unsuccessfully. Another ancient proof was based upon Eucl. 1, 11; Simson also used this proposition for the purpose, but his proof is in reality a *petitio principii.* The proposition is really required as early as 1, 1 and 1, 4. It is best therefore to make it a postulate.

POSTULATE 3.

παντὶ κέντρῳ καὶ διαστήματι, 'with every centre and distance,' meaning 'with *any* centre and distance' (cf. Note on Post. 1). διάστημα means 'distance' quite generally, as well as distance in the sense of the three *dimensions*. The word was regularly used, in describing the construction of a circle, to express the length of the radius, the idea being that all the points on the circumference of the circle are at the same *distance* from the centre. The Greeks had no word for *radius*; if they had to speak of radii, they called them 'the (straight lines drawn) from the centre,' αἱ ἐκ τοῦ κέντρου (sc. ἀγόμεναι γραμμαί).

κύκλον γράφεσθαι. In this Postulate the verb is in the passive: 'that a circle can be (literally 'is') drawn.' Proclus however has the active aorist (γράψαι) here also.

POSTULATE 4.

'That all right angles are equal to one another' (acc. and infin. after ᾐτήσθω).

In other words, a right angle is an angle of a definite size which is always the same wherever it is placed, is in fact a determinate magnitude by which other magnitudes of the same kind (acute and obtuse angles) can be measured. The statement implies the *invariability of figures*. If we attempted to prove the Postulate, we should have to use in some form or other the method of *application* of one figure to another; that is, we should have to assume the invariability of figures as an antecedent postulate. Euclid preferred to assert directly, as a postulate, that all right angles are equal; moreover it was

absolutely necessary from his point of view to put
it before Post. 5, because the condition in that
Postulate about two angles being together less than
two right angles would be useless unless it were first
made clear that right angles are angles of determinate and invariable size.

As above indicated, the truth of the Postulate
cannot at this stage be *proved* except by assuming
other postulates in place of it. One method is that
of the modern Italian school, who deduce the fact
that all right angles are equal from the equivalent
fact that all '*flat*' angles (so-called) are equal, which
is either itself assumed as a postulate or deduced
from other postulates. (A '*flat*' angle is of course
equal to two right angles, being the 'angle' which
two parts AC, CB of one straight line make with one

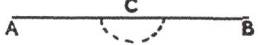

another. If we turn CA about C as centre until it
takes the position of CB, this represents a certain
definite amount of turning; a 'flat' angle therefore
represents a definite amount of turning, and a right
angle represents exactly half that amount of turning.)

POSTULATE 5.

This is the famous *Postulate* (often inappropri-
ately called the *Axiom*) *of Parallels*. It is well known
that the Euclidean theory of parallels cannot be
established except by assuming this or some other

equivalent postulate. It is clear from passages of Aristotle that in his time the theory of parallels was not grounded on a recognised postulate, but contained some *petitio principii*. Hence we may conclude that Euclid's epoch-making Postulate was first formulated by Euclid himself; and when we consider the countless attempts made through more than twenty centuries to *prove* the Postulate, many of them by geometers of great ability, we cannot but admire the genius of the man who first definitely concluded that a hypothesis such as this, which he found necessary to the validity of his whole system of geometry, was really indemonstrable.

The long sentence in which the Postulate is stated consists of the following parts: (1) a conditional clause ἐὰν...ποιῇ, (2) the accusative and infinitive ἐκβαλλομένας...συμπίπτειν governed by the original ἠτήσθω, 'let it be postulated,' (3) a relative clause ἐφ' ἃ μέρη ... ἐλάσσονες qualifying συμπίπτειν, 'meet,' and showing *in which of the two directions* the two straight lines referred to in the first words of the Postulate must meet.

The order of words in the conditional clause is ἐὰν εὐθεῖα (the subject of the clause) ἐμπίπτουσα εἰς ('falling on') δύο εὐθείας ('two straight lines') ποιῇ ('make') τὰς ἐντὸς καὶ ἐπὶ τὰ αὐτὰ μέρη γωνίας ('the angles [which are] interior and on the same side,' lit. 'towards the same parts,' i.e. regions or direction) ἐλάσσονας δύο ὀρθῶν ('less than two right angles').

We next come to the accusative and infinitive,
where the words are to be translated in this order:
τὰς δύο εὐθείας ('that the two straight lines') ἐκβαλλομένας ἐπ' ἄπειρον ('[when] produced without
limit,' cf. εἰς ἄπειρον in Def. 23 and note thereon)
συμπίπτειν ('meet,' lit. 'fall together').

Lastly, ἐφ' ἃ μέρη = ἐπὶ τὰ μέρη ἐφ' ἃ, 'on that
side on which' or 'in the direction in which,' the
whole clause ἐφ' ἃ...ἐλάσσονες meaning 'on that
side on which the angles less than two right angles
are.'

The meaning of the Postulate will easily be seen
by means of a figure. *AB* is the straight line which

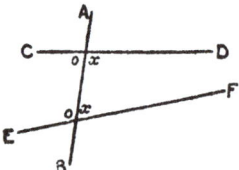

meets two other straight lines *CD, EF*. Then the
angles marked *x* are 'interior' angles 'on (one and)
the same side of' *AB*; the angles marked *o* are
'interior' angles on the other side of *AB*; both
these pairs of angles are pairs of angles which *AB*
'makes' with *CD, EF*; 'interior' angles are angles
'inside' (i.e. in the space between *CD* and *EF*).
Now suppose that, as in the figure, the sum of the
two angles marked *x* is less than two right angles.
Then the Postulate asserts that *CD, EF* will meet

on that side of AB on which the two angles marked x are situated; that is, CD, EF will meet if produced far enough on that side, i.e. if produced far enough beyond D and F respectively.

The Postulate is used in I, 29 as a means of proving that, if two straight lines (as CD, EF) are parallel, and AB meets them both, then the sum of the two angles marked x is *equal* to two right angles (and so is the sum of the two angles marked o).

THE AXIOMS OR COMMON NOTIONS

AFTER the Postulates come the Axioms or, as Euclid calls them, 'Common Notions' (κοιναὶ ἔννοιαι). These are certain self-evident truths, which cannot be proved but nevertheless have (as Aristotle says) to be assumed by any one who is to learn anything at all. They are not confined to geometry or to any one science, but are true for all alike.

Only five of the 'Common Notions' appearing in the MSS. are certainly genuine. Others were interpolated after Euclid's time.

COMMON NOTION I.

τὰ τῷ αὐτῷ ἴσα (SC. ὄντα). 'Things (which are) equal to the same (thing)'; καὶ, 'also.'

Although Aristotle had emphasised the futility

of any attempt to *prove* the axioms, it appears from Proclus that no less a person than Apollonius of Perga, ' the great geometer,' as he was called, actually made the attempt. Proclus gives as an example his attempted proof of Axiom 1. 'Let *A* be equal to *B*, and *B* to *C*; I say (says Apollonius) that *A* is also equal to *C*. For since *A* is equal to *B*, *it occupies the same space with it*; and, since *B* is equal to *C*, *it occupies the same space with it*. Therefore *A* occupies the same space with *C* [and is therefore equal to it].' The 'proof' (it will be observed) assumes two things (1) that things which 'occupy the same space' are equal to one another, and (2) that things which occupy the same space with one and the same thing occupy the same space with one another; in fact Apollonius explains the obvious by something much more obscure. Moreover the 'proof' is partial, and not general, for it is not all equal things that 'occupy space' at all.

COMMON NOTION 2.

προστεθῇ, aorist subjunctive passive of προστιθέναι, the regular word for 'to add.' 'If equals be added to equals, the wholes are equal.'

COMMON NOTION 3.

ἀφαιρεθῇ, aor. subjunctive passive of ἀφαιρεῖν, 'to take away,' 'to subtract.'

τὰ καταλειπόμενα, 'the remainders' (literally 'the things left behind').

This axiom, differing from the preceding in that the equals are subtracted fròm equals instead of being added to equals, is the axiom which is so favourite an illustration with Aristotle.

The MSS. have, after the first three Common Notions, the following four, the genuineness of which is more than doubtful.

'4. If equals be added to unequals, the wholes are unequal.

5. If equals be subtracted from unequals, the remainders are unequal.

6. Things which are double of the same thing are equal to one another.

7. Things which are halves of the same thing are equal to one another.'

They are really unnecessary and, in view of the principle that axioms ought not to be multiplied, should be omitted.

COMMON NOTION 4 [7].

The word ἐφαρμόζειν, as a geometrical term, has a different meaning according as it is used in the active or in the passive. In the passive, ἐφαρμόζεσθαι, it means 'to be applied to,' without any implication that the applied figure will exactly fit, or coincide with, the figure to which it is applied. On the other hand, the active ἐφαρμόζειν is used intransitively and means 'to fit exactly,' 'to coincide with.' In Euclid and Archimedes ἐφαρμόζειν is constructed with ἐπί and the accusative.

τὰ ἐφαρμόζοντα ἐπ' ἄλληλα, 'things which coincide with one another' (sc. when one is applied to, or laid on top of the other so as, if possible, to fit all over).

This axiom is distinctly geometrical in character and therefore hardly proper to be included among *Common Notions*. It is in fact a definition of geometrical equality more or less sufficient, but not a real axiom. This criterion of equality is required in 1, 4, but the axiom is not there referred to; Euclid says simply 'The base *BC* will coincide with the base *EF*, *and will be equal to it.*' That is to say, the axiom is *tacitly* assumed. I am inclined to think therefore that it is more likely than not to be an interpolation.

Whoever formulated this Common Notion, it seems to be intended to assert that *superposition* is a legitimate way of proving the equality of two figures which have the necessary parts respectively equal, or, in other words, to serve as an *axiom of congruence*.

It seems clear, however, that Euclid disliked the method of superposition, and avoided it wherever he could. He probably found the method handed down by tradition (if Thales proved that a circle is bisected by its diameter, he could hardly have done so by any method other than that of superposition), and he followed it, in the few cases where he does so, only because he had not been able to see his way to a satisfactory alternative.

COMMON NOTION 5 [8].

'The whole is greater than the part.'

In 1, 6, where this Common Notion might, if genuine, have been referred to, Euclid uses a different expression: 'the triangle *DCB* will be equal to the triangle *ACB*, *the less to the greater: which is absurd.*' It seems probable therefore that this axiom too was interpolated. It seems to be an abstraction or generalisation substituted for an immediate inference from a particular geometrical figure, but it takes the form of a sort of definition of whole and part.

After this Common Notion the editions up to Heiberg's generally included the interpolated axiom 'Two straight lines cannot enclose a space,' and after this the Postulates 4 and 5, which were classed as Axioms 11 and 12.

THE PROPOSITIONS

THE propositions in Euclid's *Elements* are of two kinds, known as theorems and problems. The first three in Book I are problems, the fourth begins a series of theorems. The Book contains in all 14 problems and 34 theorems.

In ancient times there was controversy as to the proper meaning of the terms, some (e.g. Speusippus and Amphinomus) holding that both kinds of proposition are alike theorems, and others (among whom was Menaechmus, the discoverer of the conic sections) maintaining that both are alike problems. So far as etymology is concerned, both views are possible. θεώρημα, *theorem*, means that which is the object of *investigation* (θεωρεῖν), and this would equally cover a problem; πρόβλημα, *problem*, means something *propounded* (προβάλλειν), which again would cover a theorem. Euclid himself does not make any distinction except that at the end of problems he has the words ὅπερ ἔδει ποιῆσαι, meaning '(which is) what it was required to do' (our *Q.E.F.*, short for *quod erat faciendum*), and at the end of theorems the different words ὅπερ ἔδει δεῖξαι '(which is) what it was required to prove' (our *Q.E.D.*, short for *quod erat demonstrandum*). These expressions indicate the distinction usually drawn. The object of a theorem is to *demonstrate* something, that of a problem to *do* or to *construct* something, although even in problems a demonstration follows, which is brought in for the purpose of confirming the construction by showing that what was required has actually been done.

In a translation from the Greek into English it is convenient to use, for the points in the figures, Roman capital letters instead of Greek, because not only are the Greek capitals rather more troublesome to write, but the parts of the figure are rather more easily grasped by means of the letters which are more familiar to us. The natural equivalents (following the proper order) are as follows:

Α Β Γ Δ Ε Ζ Η Θ Κ Λ Μ Ν Ξ Ο Π Ρ Σ Τ Υ Φ Χ Ψ Ω
= *A B C D E F G H K L M N O P Q R S T U V W X Y;*

and, except in quotations from the Greek text, I shall use the Roman letters in the following notes, in accordance with this fixed arrangement.

PROPOSITION I.

l. 1. ἐπί, 'upon.'

πεπερασμένης (perf. pass. participle of περαίνειν) means 'limited' or 'terminated'; that is, the given straight line is the straight line included between two terminal points. The word is usually translated 'finite,' and there is no objection to this.

Observe that, where we say 'on *a* given finite straight line,' the Greek usage has the more graphic expression 'on *the* given finite straight line,' i.e. the particular finite straight line which we have chosen to take.

2. συστήσασθαι, aorist infinitive middle from συνιστάναι, 'to *construct*' (literally 'set up together'), is the regular word for setting up or constructing a figure on a given base. It is best translated 'construct.' Here we are asked to construct an equilateral triangle on a given finite straight line *as base.*

3. ἔστω ἡ δοθεῖσα εὐθεῖα πεπερασμένη ἡ ΑΒ. The subject of this sentence is ἡ δοθεῖσα εὐθεῖα πεπερασμένη, the proper translation being therefore 'Let the given finite straight line be the (straight line)

AB.' Nevertheless it is best to translate as if ἡ AB were the subject 'Let *AB* be the given finite straight line.' The reason will appear when we come to such sentences as the following (in 1, 2): ἔστω τὸ μὲν δοθὲν σημεῖον τὸ A, ἡ δὲ δοθεῖσα εὐθεῖα ἡ BΓ, literally 'Let the given point be *A* and the given straight line *BC*,' which is awkward in consequence of the omission of the verb in the second clause. It is better therefore to say 'Let *A* be the given point and *BC* the given straight line.'

4. δεῖ δή, 'Thus it is required,' introduces the more specific statement of what it is required to do, with reference to the particular data (here the given straight line *AB*).

6. 'With centre *A* and distance *AB*' follows the phraseology of Post. 3 about drawing a circle. κύκλος γεγράφθω ὁ BΓΔ, literally 'let a circle *have been* described, (namely) the (circle) *BCD*.' I have already (note on Postulates, p. 142) called attention to this elegant and expressive use of the *perfect* imperative passive. In English there is no practicable alternative to translating the present phrase thus, 'let the circle *BCD* be described.'

9. ἀπὸ τοῦ Γ σημείου, καθ᾽ ὃ τέμνουσιν ἀλλήλους οἱ κύκλοι, 'from the point *C* in (lit. *at*) which the circles cut one another.' It ought to be proved that the two circles do in fact cut one another; Euclid however assumes this without proof. It is true that we can without difficulty satisfy ourselves that the circles do meet, and in *two* points, but

only by means of some postulate other than those
formulated by Euclid. There is therefore here a
real defect in Euclid's exposition.

9–11. ἀπὸ τοῦ Γ σημείου...ἐπὶ τὰ Α, Β σημεῖα ἐπεζεύχθωσαν εὐθεῖαι αἱ ΓΑ, ΓΒ, 'from the point C...
let the straight lines CA, CB [literally 'straight lines
(namely) CA, CB'] be joined to the points A, B';
a longer and more careful expression for drawing
the straight line joining two points, which gives way
later (I, 5 and afterwards) to the shortened expression 'let the straight line FC be joined' (ἐπεζεύχθω ἡ ΖΓ), the regular form in Greek geometry.
ἐπεζεύχθωσαν, *perfect* imperative passive, as usual
(ἐπιζευγνύναι).

12. καὶ = 'Then,' introduces the proof of the
correctness of the construction.

16. ἑκατέρα τῶν ΓΑ, ΓΒ. In English we have
to insert the words 'straight lines': 'each of the
straight lines CA, CB.' So (18) αἱ τρεῖς αἱ ΓΑ,
ΑΒ, ΒΓ = 'the three straight lines CA, AB, BC.'

ἄρα is the particle generally used in demonstrations where we say 'Therefore.' The force of it is
'So, then, (we may infer that).'

This first proposition gives a convenient opportunity of distinguishing the recognised formal
divisions of a proposition in Greek geometry.
Though these are best illustrated from Euclid's
propositions, they did not originate with him: on
the contrary, the form which his propositions took
was already traditional. Proclus, in his Commentary

on Book I, observes that every problem and every theorem which is complete, with all its parts perfect, purports to contain the following constituent parts.

(1) The πρότασις, 'proposition,' i.e. *'enunciation,'* which states the problem to be solved, or the theorem to be proved, in general terms. Cf. the enunciation of this proposition 'On a given finite straight line to construct an equilateral triangle.'

(2) The ἔκθεσις, *'setting-out'* (ἐκτιθέναι). This sets out the data in a concrete form ready for use in the subsequent stages. In I, I the 'setting-out' consists of the words 'Let *AB* be the given finite straight line.' The straight line *AB* is the one datum in this case.

(3) The διορισμός, literally the 'delimitation' or 'definition' (but not definition in the ordinary sense, the Greek word for which is ὅρος). By this is meant the specification or particular statement of what the theorem or problem requires us to prove or to construct, not in general terms (as in the enunciation) but with reference to the particular concrete data which are shown in a diagram; the object of the διορισμός is then to fix the attention better, by visualisation, as it were. In I, I, it consists of the words 'Thus it is required to construct an equilateral triangle on the straight line *AB*.' It is regularly introduced by the words δεῖ δή, 'Thus it is required.'

(4) The κατασκευή, *apparatus* or *construction*. In the particular case of I, I the *construction* consists of

the drawing of the two circles and of two lines joining pairs of points.

(5) The ἀπόδειξις, *proof* or *demonstration*.

(6) The συμπέρασμα, *conclusion*, which reverts to the enunciation and shows that the problem has been solved or the theorem proved. In 1, 1 it consists of the words 'Therefore the triangle *ABC* is equilateral, and it has been constructed on the given finite straight line *AB*,' and in other problems it takes a similar form. In theorems the conclusion generally repeats the enunciation word for word. The conclusion is immediately followed in the case of problems by the words ὅπερ ἔδει ποιῆσαι, '(which is) what it was required to do,' *quod erat faciendum (Q.E.F.)*, and in the case of theorems by the words ὅπερ ἔδει δεῖξαι, '(which is) what it was required to prove,' *quod erat demonstrandum (Q.E.D.)*.

In particular cases some of the above divisions are not found; but all propositions contain at least three of them, the *enunciation*, the *proof*, and the *conclusion*.

The word διορισμός (cf. 3 above) has another and a more important sense in relation to a *problem*. Problems may be possible or impossible of solution according to the nature and relations of the data. This is best explained by an example. We learn from Eucl. 1, 20 that any two sides of a triangle are together greater than the third. Hence, when we are asked to construct a triangle with sides equal to three given straight lines respectively (1, 22), it is necessary to be satisfied beforehand that, with the given data, a solution is possible or, in other words, that the three given straight lines are such as *can* form a triangle; and by 1, 20 this can only be when any two of the straight lines are together greater than the third. Accordingly Euclid adds to the enunciation of 1, 22 the words 'Thus it is necessary (δεῖ δή) that two of the straight lines taken together in any manner should be greater than the remaining straight line.' This is a formal statement of the necessary condition which must be fulfilled if the solution of the problem is to be possible; and this is what the Greeks meant by διορισμός. διορίζειν = δι + ὁρίζειν, literally 'to set boundaries (ὅρος) between,' i.e. to make a demarcation between the cases in which the solution is possible and those in which it is not. It is difficult to find an English word that is a proper equivalent. If we say *definition* we have to remember that it is a different thing from a definition in the ordinary sense; 'delimitation' might perhaps serve, but it is probably better to abandon the attempt to be literal and to say 'limiting condition.'

PROPOSITION 2.

1. πρὸς τῷ δοθέντι σημείῳ, 'at the given point,' in accordance with the Greek usage, whereas we say 'at a given point' and should so translate.

2. θέσθαι, 'to place' (aor. inf. mid. of τιθέναι). To place a straight line *at* a point means to place it so

that one of its extremities is at the point. We may,
in order to make this clear, translate 'To place with
its extremity at a given point a straight line equal
to a given straight line.'

7. συνεστάτω, perf. imperative of συνιστάναι, with
passive sense, 'let the equilateral triangle DAB be
constructed' (lit. 'have been constructed').

8. ἐκβεβλήσθωσαν ἐπ᾽ εὐθείας ταῖς ΔA, ΔB
εὐθεῖαι αἱ AE, BZ, 'let the straight lines AE, BF be
produced in a straight line with DA, DB.' Observe
that the straight lines here said to be 'produced'
are not the original straight lines DA, DB but the
produced portions of them. Euclid has this longer
description of the operation of producing a straight
line in 1, 5 also; it is not till 1, 16 that we find him
using the shorter form familiar to us, 'let BC be
produced to D.'

16. ὧν ἡ ΔA τῇ ΔB ἴση ἐστίν. It is better to begin
a new sentence here and to combine this clause with
the next in one sentence. 'And in these (ὧν) DA
is equal to DB; therefore the remainder AL is equal
to the remainder BG.' The expression λοιπὴ ἡ AΛ
means literally 'AL remaining' or 'which remains';
but it is hardly possible to say in English 'AL which
remains is equal to BG which remains'; it is better
to translate λοιπὴ ἡ AΛ by 'the remainder AL':
'the remainder AL is equal to the remainder BG.'

The reason for the insertion of this proposition is
that Euclid does not regard it as legitimate to
suppose that we can *take up* a straight line of given

length from one position and place it in another
(e.g. with its extremity at a given point). He would
regard it as ungeometrical to *carry* a distance by
means of a ruler with marks on it showing the distance, or by means of an opened pair of compasses.
As De Morgan says, we are not allowed to draw a
circle with a compass-carried distance; it is as if we
had to suppose that the compasses would close of
themselves the moment they cease to touch the
paper.

There are several possible *cases* of this proposition.
The relative positions of the point and the straight
line may vary, and we may describe the equilateral
triangle *DAB* on either side of *AB*. According to
these differences the figure of the proposition takes
slightly different forms; the proof is however,
mutatis mutandis, the same. Two other figures are
added to show this.

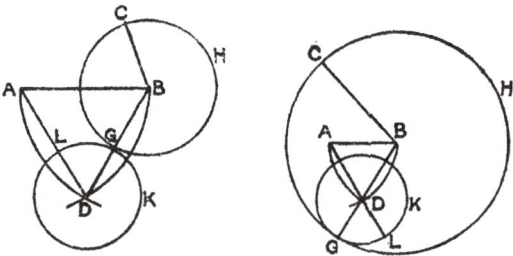

In one of them it is not necessary to produce the
side of the equilateral triangle and *AL*, *BG* are the

differences between *AD*, *DL* and *BD*, *DG* instead of the differences between *DL*, *AD* and *DG*, *BD* respectively; while in the other figure *AL* and *BG* are the *sums* instead of the *differences* of *DL*, *AD* and *DG*, *BD* respectively. The pupil can work out the proof in these or other cases for himself.

The Greek commentators revelled in distinctions of cases; the classical Greek geometers, and notably Euclid, were in the habit of giving one case only, for choice the most difficult, leaving the reader to think out the rest for himself. Where there was a real disparity between the cases, sufficient to involve a substantial difference in the proofs, the practice was to give separate enunciations and proofs altogether. This is seen in the *Conics* and other treatises of Apollonius.

PROPOSITION 3.

1. We may, if we please, translate δύο δοθεισῶν εὐθειῶν ἀνίσων as if it were a genitive absolute, 'given two unequal straight lines,' although the genitive may equally be taken as the partitive genitive after ἀπὸ τῆς μείζονος, 'from the greater of two given straight lines.'

2. The order of the words is ἀφελεῖν εὐθεῖαν ἴσην τῇ ἐλάσσονι, 'to cut off a straight line equal to the less.'

7. κείσθω. κεῖσθαι is here, as usual, employed as the passive of τιθέναι, 'let (*AD*) *be placed*' (cf. Prop. 2).

PROPOSITION 4.

2. ἑκατέραν ἑκατέρᾳ. The traditional translation is 'each to each,' which however is unsatisfactory because it might be inferred from it that all the four sides mentioned (two in each triangle) are equal, whereas what is meant is that one of the two sides in the first triangle is equal to one of the two sides in the second triangle, and that the second of the two sides in the first triangle is equal to the second of the two sides in the second. Translate by 'respectively' here and always. 'If two triangles have two sides equal to two sides respectively,' i.e. if two sides in the one triangle be equal respectively to two sides in the other. I have translated τὰς δύο πλευρὰς by 'two sides' (without the definite article) because our English practice substitutes the indefinite article for the definite in such a case.

2–4. τὴν γωνίαν τὴν ὑπὸ τῶν ἴσων εὐθειῶν περιεχομένην, 'the angle contained by the equal straight lines.' This means the angle included in the one triangle by the two sides which are *respectively* equal to the two sides in the other. With τῇ γωνίᾳ we should, strictly speaking, understand a similar expression τῇ ὑπὸ τῶν ἴσων εὐθειῶν περιεχομένῃ. We may translate 'and have the angles contained by the equal straight lines equal.' Euclid no doubt says 'straight lines' rather than 'sides' here in order to follow scrupulously at the outset the language of Def. 9.

4. καὶ τὴν βάσιν. The apodosis begins here. 'They will also have the base equal to the base.'

7, 8. ὑφ' ἃς αἱ ἴσαι πλευραὶ ὑποτείνουσιν : ὑφ' ἃς qualifies αἱ λοιπαὶ γωνίαι. ὑποτείνειν means literally 'stretch underneath,' and with ὑπό and acc. it means 'subtend' in the sense of 'being opposite to.' 'The remaining angles will be equal to the remaining angles respectively, namely those which the equal sides subtend.' Cf. ἡ ὑποτείνουσα, the regular word for the *hypotenuse* of a right-angled triangle, which subtends a particular angle in the right-angled triangle, namely the right angle.

12. ἡ ὑπὸ ΒΑΓ γωνία, the regular Greek expression (with or without γωνία) for 'the angle *BAC*.' The full expression would be ἡ ὑπὸ τῶν ΒΑ, ΑΓ περιεχομένη γωνία, 'the angle contained by the (straight lines) *BA*, *AC*.' But it was a common practice of Greek geometers, e.g. Archimedes and Apollonius (though not apparently of Euclid himself), to write αἱ ΒΑΓ for αἱ ΒΑ, ΑΓ, 'the straight lines *BA*, *AC*.' Thus, when περιεχομένη was dropped, the expression for the angle *BAC* became, first ἡ ὑπὸ τῶν ΒΑΓ γωνία, then ἡ ὑπὸ ΒΑΓ γωνία, and finally ἡ ὑπὸ ΒΑΓ, without γωνία, as regularly found in Euclid.

13. Observe that the 'particular statement' of the proposition is, in the case of theorems, introduced by the word λέγω 'I say' (that...).

19. ἐφαρμοζομένου τοῦ τριγώνου, 'if the triangle be *applied* (to)': this is the regular meaning of the *passive* ἐφαρμόζεσθαι.

21. ἐφαρμόσει, 'will *coincide* (with)' or 'exactly fit'; this is the meaning of the *active* forms of ἐφαρ-

μόζειν, which are used intransitively (cf. note on Common Notion 4 above, p. 153). This distinction between the active and the passive of the verb ἐφαρμόζειν is well brought out in this proposition.

28. 'Hence the base *BC* will coincide with the base *EF*, (33) and will be equal to it.' The coincidence of *BC* with *EF* is immediately inferred from the fact that the ends of the two straight lines coincide respectively. Euclid uses the fact, implied in Post. 1, that there is only one straight line joining two points. Some commentator, not deeming the argument sufficiently clear, inserted an explanation in support of the inference that *BC* coincides with *EF*: (29–33) 'For if, when *B* coincides with *E* and *C* with *F*, the base *BC* does not coincide with the base *EF*, two straight lines will enclose a space: which is impossible. Therefore the base *BC* will coincide with *EF*,' introducing in this way the interpolated and unnecessary axiom that 'two straight lines cannot enclose a space.'

39–45. Where Euclid, in his conclusion, simply (as here) repeats the enunciation word for word, we can simply write 'Therefore etc.'

PROPOSITION 5.

1. αἱ πρὸς τῇ βάσει γωνίαι, 'the angles *at* the base,' meaning 'adjacent to the base,' i.e. the angles formed with the base by the two equal sides respectively.

2. προσεκβληθεισῶν τῶν ἴσων εὐθειῶν, genitive

absolute, 'when' or 'if the equal straight lines are produced further' (πρός, 'in addition'), the 'equal straight lines' meaning the equal *sides*.

10. εἰλήφθω γὰρ ἐπὶ τῆς ΒΔ τυχὸν σημεῖον τὸ Ζ. τυχόν is the aorist participle of τυγχάνειν, so that τυχὸν σημεῖον is 'any point *as it may happen*,' 'a chance point': 'For let a point F be taken at random on BD.'

15. γωνίαν κοινὴν περιέχουσι τὴν ὑπὸ ΖΑΗ, 'they (i.e. the two sides in the two triangles FAC, GAB respectively) contain a common angle, the angle FAG.' κοινὴν means 'common to both triangles,' and its force is here predicative, as if we were to say 'the angle FAG which they contain is common (to both triangles),' and therefore *equal* in both triangles.

26. βάσις αὐτῶν. αὐτῶν here evidently means the *angles* just mentioned: 'the base BC is common to them.' The triangles BFC, CGB again, having a common base, have their bases *equal*.

35. εἰσὶ πρὸς τῇ βάσει, 'are at the base.' The subject is the angles ABC, ACB. The angles ABC, ACB 'are (the angles) at the base.'

This, being the first proposition which has to be closely argued, is no doubt difficult for a beginner: hence the name given to it, *pons asinorum*. The points to remember are that the proposition 1, 4 has to be applied, *first* to the large triangles FAC, GAB, and *secondly* to the small triangles beyond the base, namely FBC, GCB. The second result of the proposition, the equality of the angles beyond or under the base, follows from the equality in all respects of the latter pair of triangles. And, lastly, the equality of the angles *at* the base is proved by subtracting equals from equals (namely equal angles in the triangles FBC, GCB from equal angles in the triangles FAC, GAB).

It is to be observed that the pairs of triangles as they appear in the figure are not *superposable* as the triangles in I, 4 are: they are *symmetrically* equal rather than equal in the sense of I, 4, though they would be equal in that sense if one of the two were regarded as turned over or seen from the back.

A much shorter proof of the theorem of I, 5 was given by Pappus. It requires no 'construction': it suffices simply to draw the triangle ABC.

'Let us,' says Pappus, 'conceive the one triangle [ABC] as two triangles [i.e. as two triangles ABC and ACB respectively] and let us argue in this way.

Since AB is equal to AC, and AC to AB, the two sides AB, AC are equal to the two sides AC, AB.

And the angle BAC is equal to the angle CAB, for it is the same.

Therefore all the corresponding parts (in the triangles) are equal, namely

$$BC \text{ to } CB,$$

the triangle ABC to the triangle ACB, and the angle ACB to the angle ABC (for these are the angles subtended by the equal sides AB, AC).

Therefore in isosceles triangles the angles at the base are equal.'

If there is any difficulty in visualising the one triangle as two, we may regard the second triangle *ACB* as the first triangle *ABC seen from the back.*

We are told by Proclus that the first to discover the truth of this proposition was Thales, who however spoke of the equal angles as 'similar' angles.

Aristotle refers to a proof of the theorem of 1, 5 which is quite different from Euclid's. Aristotle no doubt found this proof in the text-book of Elements in use in his time, which was probably that of Theudius of Magnesia. We infer that Euclid's proof was his own, and in fact that he arranged his book from the very beginning on a plan altogether different from that followed by his predecessors.

PROPOSITION 6.

2. αἱ ὑπὸ τὰς ἴσας γωνίας ὑποτείνουσαι πλευραί: as usual, ' the sides *subtending* [i.e. opposite to] the equal angles.'

15. 'The triangle *DBC* will be equal to the triangle *ACB, the less to the greater*: which is absurd.' Euclid assumes that, because *D* is between *A* and *B*, the triangle *DBC* is less than the triangle *ABC*. To prove this theoretically would require some postulate; but Euclid is satisfied to make the inference directly from the look of the figure.

As regards this proposition we have to observe (1) that it is what is called the *converse* (ἀντίστροφος) of the preceding Prop. 5, and (2) that the method of proof by *reductio ad absurdum* is here used by Euclid for the first time.

(1) *Conversion* (in Greek ἀντιστροφή, 'turning the opposite way,' from ἀντί and στρέφειν) may be either *complete* or *partial*. This proposition is a case of *complete* conversion, conversion, that is, where the hypothesis and the conclusion of a theorem change places exactly. The hypothesis of 1, 5, the equality of the two *sides*, is the conclusion of 1, 6, and the conclusion of 1, 5, the equality of the two *angles*, is the hypothesis of 1, 6. We shall find a case of *partial* conversion when we come to 1, 8, which is the *partial* converse of 1, 4.

(2) The method of proof by *reductio ad absurdum* is not peculiar to geometry. It is described by Aristotle as a logical method of proof (*Analytica priora*, Book 1); he calls it alternatively ἡ εἰς τὸ ἀδύνατον ἀπαγωγή, 'reduction to the impossible,' or ἡ διὰ τοῦ ἀδυνάτου δεῖξις or ἀπόδειξις, 'proof by the impossible,' or ἡ εἰς τὸ ἀδύνατον ἄγουσα ἀπόδειξις, 'proof leading to the impossible.'

'Proof leading to the impossible,' says Aristotle, 'differs from the direct (δεικτικῆς, demonstrative) in that it assumes what it desires to destroy [namely the hypothesis that the proposition which we seek to prove is false], and then reduces it to something which is admittedly false.' If we arrive at a conclusion admittedly false, this shows that our hypothesis was false, and therefore that the original proposition which we desired to prove, and which that hypothesis contradicted, cannot but be true. So in this proposition, where we have to prove that the sides are equal, we begin by assuming that they are *un*-equal and show that this hypothesis leads to a conclusion which is obviously false. It follows that the hypothesis is false, that is, that the sides are *not* unequal as we supposed, and therefore must be equal. Proclus also has a good description of *reductio ad absurdum*: 'Every *reductio ad absurdum*,' he says, 'assumes what conflicts with the desired result, then, using that as a basis, proceeds until it arrives at an admitted absurdity and, by thus destroying the hypothesis, establishes the result originally desired.'

PROPOSITION 7.

1–5. The enunciation of this proposition is very difficult to render intelligibly, because the wording is extremely condensed. The fact is that the phraseology is traditional and would convey more to the contemporary Greek than is actually contained in the words; for we find the identical phrase οὐ συσταθήσονται πρὸς ἄλλο καὶ ἄλλο σημεῖον used with the same meaning in Aristotle.

Just as συστήσασθαι ἐπὶ is used of *constructing* a triangle on a given base (cf. 1, 1), so it is used of setting up or constructing two straight lines from the ends of a straight line so as to meet in a point. Let us now analyse the sentence. ἐπὶ τῆς αὐτῆς εὐθείας means 'on the same straight line [i.e. a given straight line] (as base).' The subject of the sentence is ἄλλαι δύο εὐθεῖαι, 'two other straight lines,' and the words qualifying it are ἴσαι δύο ταῖς αὐταῖς εὐθείαις ἑκατέρα ἑκατέρᾳ, 'equal to the same two straight lines (i.e. to two given straight lines) respectively.' οὐ συσταθήσονται, 'cannot be constructed' (literally 'shall not be constructed'), πρὸς ἄλλῳ καὶ ἄλλῳ σημείῳ, 'to different points' (literally 'to another and another point'), the meaning being that 'the same two straight lines' are constructed to *one* point and the 'two other straight lines' are supposed to be constructed to *another* point; ἐπὶ τὰ αὐτὰ μέρη, 'on the same side' (sc. of the given base). The last expression τὰ αὐτὰ πέρατα ἔχουσαι ταῖς ἐξ ἀρχῆς εὐθείαις qualifies ἄλλαι δύο εὐθεῖαι as predicate and means 'having (or 'if they have') the same extremities respectively as the original straight lines'; i.e. the lines which are supposed to be respectively equal are those starting from the same end of the given base.

The nearest possible approach to a literal translation serving to convey the full meaning seems to be the following.

'Given two straight lines [constructed on a given straight line from its extremities and meeting at a point], there cannot be constructed on the same straight line two other straight lines meeting in another point on the same side [of it] and equal to the former two respectively, namely each to that which has the same extremity with it.'

Simson altered the wording to the following

which expresses the same thing more simply: 'On the same base and on the same side of it there cannot be two triangles that have their sides which are terminated at one extremity of the base equal to one another and likewise those terminated at the other extremity.'

The whole thing is made intelligible by the *setting-out*, with reference to the figure of the proposition.

15. μείζων ἄρα ἡ ὑπὸ ΑΔΓ τῆς ὑπὸ ΔΓΒ, 'therefore the angle *ADC* is greater than the angle *DCB*.' This inference would be more obvious if, immediately before it, we inserted, as an intermediate step, 'But the angle *ACD* is greater than the angle *DCB*.'

16. πολλῷ μείζων, 'much greater' (literally 'greater by much'), the usual expression in Greek geometry for '*a fortiori* greater.'

There are two possible cases of this proposition. Euclid, in accordance with the usual habit of the classical Greek geometers, only gives one case, leaving the other to be worked out by the pupil. It is

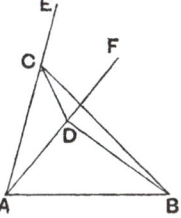

supplied by Proclus, and is shown in the accompanying figure. In this case *D*, instead of being on the side of *CB* away from *A*, is inside the triangle.

We join CD and produce AC, AD to E, F.

Then, since $AC = AD$, the triangle ACD is isosceles, and therefore [*second* part of 1, 5] the angles ECD, FDC, *under* the base, are equal.

Therefore the angle FDC, being equal to the angle ECD, is greater than the angle BCD;
therefore the angle BDC [which is greater than the angle FDC] is much greater than the angle BCD.

But since, in the triangle BCD, $BC = BD$ (by hypothesis), the angle BDC is equal to the angle BCD.   [1, 5]

But it was proved much greater: which is impossible.

Prop. 7 is, actually, only used by Euclid for the purpose of proving the following proposition, 1, 8. It has however an independent value of its own, as is easily seen. For it proves that, if, in addition to the base of a triangle, the length of the side terminated at each extremity of the base is given, only one triangle satisfying these conditions can be constructed on one and the same side of the given base. It shows therefore that the constructions of 1, 1 and 1, 22 give only *one* triangle on one and the same side of the base.

PROPOSITION 8.

12, 14 &c. ἐφαρμόζεσθαι ἐπί, 'to be applied to'; ἐφαρμόζειν ἐπί, 'to coincide with,' as usual.

20. παραλλάξουσιν, 'will fall beside them';

παραλλάττειν (παρά and ἀλλάττειν, 'to change') means 'to pass by without touching,' 'to miss,' or 'to fall awry.'

24. οὐ συνίστανται δέ, 'but they cannot be so constructed' [Prop. 7].

Prop. 8 is, as already mentioned, a partial converse of Prop. 4. The meaning of *partial* conversion will be clear from this example. The enunciation of Prop. 4 contains *two* hypotheses, (1) that two sides in two triangles are respectively equal, (2) that the included angles are equal, and the conclusion proved is (3) that the bases are equal. Prop. 8, the partial converse, takes as its two hypotheses *one* of the hypotheses of Prop. 4, namely (1), and the *conclusion* of that proposition (3), and proves as the *conclusion* the second hypothesis of Prop. 4, namely that the angles included by the equal sides are equal. More generally, the partial converse takes as hypotheses the conclusion and all the hypotheses but one of the original theorem, and deduces, as the conclusion, the remaining hypothesis of the original theorem.

An ancient alternative proof of 1, 8, which avoids the use of 1, 7, is due to Philo of Alexandria. It has the disadvantage that it is necessary to consider three cases.

Let *ABC*, *DEF* be two triangles in which the sides *AB*, *AC* are equal respectively to the two sides *DE*, *DF*, and the base *BC* is equal to the base *EF*.

Let the triangle *ABC* be taken up and placed so that *BC* coincides with *EF* but the point *A* falls at

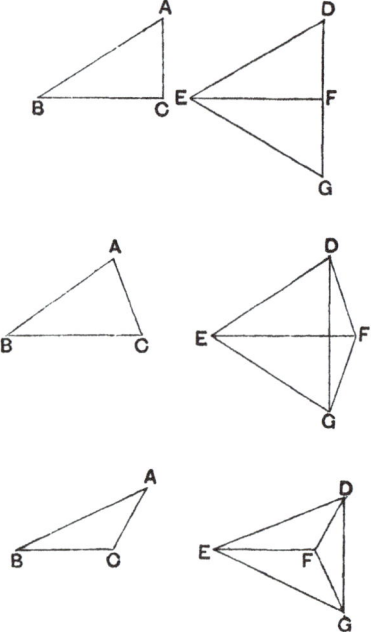

G on the opposite side of *EF* from *D* (i.e. the triangle *ABC* is *turned over* as well as moved to a new position).

Then either *DF*, *FG* are in a straight line or they are not. If they are not, join *DG*.

DG may in the latter case either meet *EF* between

E and F or may meet EF produced in either direction.

There are therefore three cases as shown in the figures.

In the first case, where DF, FG are in a straight line, we have a triangle DEG in which the sides DE, EG are equal. Therefore (by I, 5) the angle EDG is equal to the angle EGD.

In the second and third cases we have two triangles DEG, DFG which are isosceles, since, by hypothesis, DE is equal to EG, and DF to GF. Therefore, in the triangle DEG, the angles EDG, EGD are equal, and, in the triangle DFG, the angles FDG, FGD are equal. Therefore in the second figure the sum of the angles EDG, FDG is equal to the sum of the angles EGD, FGD, so that the whole angle EDF is equal to the whole angle EGF. In the third figure the *difference* between the angles EDG, FDG is equal to the *difference* between the angles EGD, FGD; that is, the angle EDF is again equal to the angle EGF.

Therefore in all the cases the angle EDF is equal to the angle EGF, which is, by hypothesis, equal to the angle BAC.

PROPOSITION 9.

1. 'To bisect a given rectilineal angle.' δίχα τεμεῖν (aor. inf. of τέμνειν), literally 'to cut in two,' means to cut *into two equal parts*, i.e. to bisect (cf. Def. 17).

4. τυχὸν σημεῖον, 'a chance point,' as usual: 'let a point *D* be taken at random on *AB*.'

8. δίχα τέτμηται, 'has been bisected.'

Observe that Euclid describes the equilateral triangle on the side of *DE* opposite to *A*, although he does not note the fact. In general it would make no difference on which side the triangle is drawn;

but, if it is drawn on the side of *DE* towards *A*, the construction would fail in one particular case, namely that in which the given angle is equal to the angle of an equilateral triangle; for then *F* would coincide with *A*, and there would be nothing to show in which direction a straight line should be drawn through *A* in order to bisect the angle *DAE*.

PROPOSITION 10.

1. 'To bisect a given finite straight line,' i.e. a straight line terminated at two given points, as distinct from an unlimited straight line.

Apollonius, we are told, bisected a straight line by a construction like that of 1, 1. With centres *A*, *B* and radii *AB*, *BA* respectively describe circles intersecting at C, D. Join CD meeting AB in E. AB is then bisected at E.

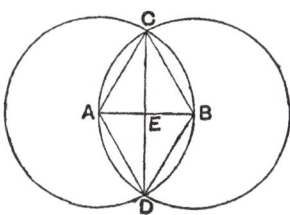

To prove this we join AC, CB, BD, DA and then prove (1) that the triangles ACD, BCD are equal in all respects (this follows from 1, 8), and therefore that the angle ACD is equal to the angle BCD, and then (2) that the triangles ACE, BCE are equal in all respects (this follows from 1, 4), and therefore that AE is equal to EB.

It will be observed that Apollonius's method is equivalent to that in use in *practical* geometry.

PROPOSITION II.

1. τῇ δοθείσῃ εὐθείᾳ, dative after πρὸς ὀρθὰς γωνίας; 'at right angles to a given straight line.'

ἀπὸ τοῦ πρὸς αὐτῇ δοθέντος σημείου, 'from a given point *on* it,' i.e. on the given straight line.

9. κείσθω τῇ ΓΔ ἴση ἡ ΓΕ, 'let *CE be made* equal to *CD*.' As usual, κεῖσθαι is used for the passive of τιθέναι, which means 'to place' but is often used in the sense of *making* (e.g. making one straight line equal to another straight line).

18, 19. ἐφεξῆς, 'adjacent,' as in Def. 10.

If we regard *CD* as making with its continuation
CE what is often called a '*flat* angle,' the construction of this proposition in effect bisects the flat
angle by means of the straight line *CF*. The construction therefore is really a particular case of that
in I, 9. The equal halves of the 'flat angle,' namely
the angles *DCF*, *FCE* are right angles in accordance
with Def. 10, and each of the right angles is seen to
be half a 'flat angle.'

If it is difficult to conceive of the 'flat angle' as
an angle, the idea of the *size* of this angle can be

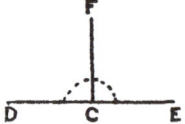

got by reflecting that, if *CD* be turned about *C* as
centre until it coincides with *CE*, the flat angle
represents twice the amount of turning that the
right angle *DCF* represents; for, in order to arrive
at *CE*, *CD* has to describe successively the *two* right
angles *DCF*, *FCE*.

PROPOSITION 12.

1. ἐπὶ τὴν δοθεῖσαν εὐθεῖαν ἄπειρον, 'to (lit.
'upon') a given unlimited straight line,' after
κάθετον . . . ἀγαγεῖν.

2. ὃ μή ἐστιν ἐπ' αὐτῆς, 'which is not on it' (sc.
the given straight line).

3. κάθετον εὐθεῖαν γραμμήν, 'a perpendicular straight line.' This is the full expression for a 'perpendicular.' κάθετος (καθιέναι) means 'let fall' or 'let down,' 'dropped,' so that the expression corresponds to our *plumb-line*. ἡ κάθετος is constantly used alone for a perpendicular, εὐθεῖα γραμμή being understood.

9. ἐπὶ τὰ ἕτερα μέρη τῆς AB εὐθείας, 'on the other side of the straight line *AB*' (lit. 'towards the other parts of the straight line *AB*'). Cf. ἐπὶ τὰ αὐτὰ μέρη, 'on the same side' in Post. 5 and Prop. 7; ἐφ' ἑκάτερα τὰ μέρη, 'in both directions' (Def. 23).

D is taken at random *on the side of AB away from C* in order that it may be certain that the circle drawn with *C* as centre and *CD* as radius will really cut *AB* in two points. Just as in 1, 1, however, we need some postulate to assure us of the fact.

Proclus tells us that this problem was first investigated by Oenopides of Chios (fifth century B.C.), who thought it useful for astronomy. Oenopides called the perpendicular κατὰ γνώμονα, '(a straight line drawn) in the manner of a gnomon' (or 'gnomon-wise'). The *gnomon* was (in its original meaning) a stick or needle placed in a vertical position for the purpose of casting shadows and so serving as a means of measuring time; hence the use of the term for a perpendicular. It is not to be supposed that Oenopides was the first to draw perpendiculars in geometrical figures. What is meant is doubtless that Oenopides was the first to give the present theoretical construction; previously, no

doubt, perpendiculars would be drawn by means of
a set-square or other equivalent mechanical means.

PROPOSITION 13.

I. σταθεῖσα (aor. pass. participle of ἱστάναι),
'set up': 'if a straight line set up on a straight line
make angles' (γωνίας ποιῇ), i.e. if the first straight
line makes (two) angles with the second straight line
on which it is set up. 'It will make either two right
angles' (i.e. each of the two angles will be a right
angle) 'or angles equal to two right angles' (i.e. it
will make unequal angles, the sum of which however
will be equal to two right angles): with ὀρθὰς and
ἴσας understand γωνίας.

II. κοινὴ προσκείσθω ἡ ὑπὸ ΕΒΔ. As usual, προσ-
κεῖσθαι is used in place of the passive of προστιθέναι,
'to add.' The phrase is very characteristic in Greek
geometry, but a literal translation is hardly practicable. The meaning is '*let* the angle *EBD be added
(so as to be) common*' (to both equals). 'Let the
common angle *EBD* be added' would clearly be
an inaccurate rendering, since the angle is not
common before it is added; the κοινὴ is *proleptic*. In
the corresponding expression for subtraction, κοινὴ
ἀφῃρήσθω (ἀφαιρεῖν), 'let the common angle be
subtracted' would be less unsatisfactory as a translation; but, as it is desirable to use corresponding
words when translating the two expressions, it is
probably best to say 'let the angle *EBD* be added
to each' (or 'subtracted *from each*'), though we may,

if we like, add '(in) common' in order to render
κοινή.

PROPOSITION 14.

1. In the conditional sentence δύο εὐθεῖαι μὴ ἐπὶ
τὰ αὐτὰ μέρη κείμεναι is the subject and goes with
ποιῶσι τὰς ἐφεξῆς γωνίας δυσὶν ὀρθαῖς ἴσας, 'make
the adjacent angles equal to two right angles.'

Observe the two meanings of πρός in the enunciation. πρός τινι εὐθείᾳ = 'with any straight line,'
which is constructed with ποιῶσι γωνίας, 'make
angles with'; καὶ (πρὸς) τῷ πρὸς αὐτῇ σημείῳ, 'and
at a point on it' (sc. the straight line).

2. μὴ ἐπὶ τὰ αὐτὰ μέρη κείμεναι, qualifying δυο
εὐθεῖαι, 'two straight lines not lying on the same
side' (of the straight line first mentioned).

The whole enunciation is: 'If with any straight
line, and at a point on it, two straight lines not
lying on the same side make the adjacent angles (together) equal to two right angles, the two straight
lines will be in a straight line with one another.'

19. ὁμοίως δὴ δείξομεν, 'similarly we can prove'...
(literally 'we shall prove,' i.e. if we please).

PROPOSITION 15.

2. κορυφή, *vertex*, so that κατὰ κορυφήν means
'by way of vertex' or 'vertically related.' We may
translate αἱ κατὰ κορυφὴν γωνίαι (the angles which
are 'by way of vertex,' sc. to one another) by 'the
vertical angles,' by which are meant the vertically
opposite angles.

This theorem, according to Eudemus, was first discovered by Thales but found its scientific demonstration in Euclid.

In one MS. and in the margin of another (not the best) a *corollary* is added. The Greek word for what we call 'corollary' is πόρισμα, meaning something 'provided' or 'found' (πορίζω). In the *Elements* the word (which we may translate by 'Porism') always has the meaning of corollary, which is an incidental result springing from the proof of a theorem or the solution of a problem, a result not directly sought but appearing, as it were, by chance without any additional labour and constituting, as Proclus says, a sort of windfall (ἕρμαιον) or *bonus* (κέρδος).

The text of the Porism is Ἐκ δὴ τούτου φανερὸν ὅτι, ἐὰν δύο εὐθεῖαι τέμνωσιν ἀλλήλας, τὰς πρὸς τῇ τομῇ γωνίας τέτρασιν ὀρθαῖς ἴσας ποιήσουσιν, 'From this it is manifest that, if two straight lines cut one another, they will make the angles at the (point of) section equal to four right angles.' τομή, 'section,' is regularly used of the *point of* section of two lines and of the *line of* section of two surfaces.

Proclus adds notes (1) that it follows from the Porism that, if any number of straight lines intersect one another at one point, the sum of all the angles so formed about the point will still be equal to four right angles (this corresponds to what is usually given in the text-books as Cor. 2), and (2) that the Porism forms the basis of the theorem which proves that only the following three (regular) polygons can fill up the whole space surrounding one point, namely the equilateral triangle, the square, and the equilateral and equiangular hexagon, a theorem which, he says, was due to the Pythagoreans.

PROPOSITION 16.

1. παντὸς τριγώνου. The Greek has, as usual, 'every triangle' where we say '*any* triangle.' The genitive is constructed with ἡ ἐκτὸς γωνία; but it is best to translate παντὸς τριγώνου by 'In any triangle.'

2. ἡ ἐκτὸς γωνία, 'the exterior angle' (literally 'the outside angle'), is the angle formed by any side and an adjacent side produced.

ἑκατέρας τῶν ἐντὸς καὶ ἀπεναντίον γωνιῶν, 'than each of the interior and opposite angles.' ἐντός and ἀπεναντίον are adverbs but are used here, like ἐκτός, adjectivally.

10. διήχθω (perfect imperative passive of διάγειν), a variation on the usual ἐκβεβλήσθω: 'let *AC* be drawn (or 'carried') through.'

PROPOSITION 17.

1. παντὸς τριγώνου, 'In any triangle,' as usual.

αἱ δύο γωνίαι. As usual, the Greek idiom uses the definite article where we say 'two angles' simply. The force of the definite article is 'the two angles that may be taken (whichever they are).'

2. πάντῃ μεταλαμβανόμεναι, 'taken together in any manner'; the meaning is 'any two angles added together.'

We may translate 'In any triangle two angles taken together in any manner are less than two right angles.'

7. ἐκτός ἐστι γωνία. ἐκτός is again used adjectivally: 'since the angle *ACD* is an exterior angle of the triangle *ABC*.'

13. ὁμοίως δὴ δείξομεν, 'similarly we can prove,' as usual.

PROPOSITION 18.

It is necessary to be clear as to the distinction between this proposition and the next, to know, that is, which comes first and is proved directly, and which comes second and is proved by *reductio ad absurdum*; and, also, in which of the two it is the greater *side* that is given and in which the greater *angle* is the datum. On the first point Todhunter has the useful note 'In order to assist the student in remembering which of these two propositions (I, 18, 19) is demonstrated directly and which indirectly, it may be observed that the order is similar to that in I, 5 and I, 6.'

On the second point it is to be noted that the datum comes first in each enunciation. Thus in I, 18 the enunciation 'In any triangle the greater side subtends the greater angle' means that we are given that one *side* is greater than another and have to prove that, of the angles which the two sides respectively subtend, the angle subtended by the greater side is the greater.

2. In this enunciation we have ὑποτείνειν ('to subtend') used with the simple accusative instead of the more usual ὑπό and accusative.

6. κείσθω τῇ AB ἴση ἡ AΔ: as usual, 'let *AD be made* equal to *AB*.'

PROPOSITION 19.

In order to get the datum (here the greater *angle*, not the greater side) first in the enunciation, we can use the passive of the verb for 'subtend' in the translation instead of the active. 'In any triangle the greater angle is subtended by the greater side.'

7, 8. ἴση γὰρ ἂν ἦν, 'would have been equal,' οὐκ ἔστι δέ, 'but it is not.'

9. οὐδὲ μήν, 'nor yet' or 'nor again.'

PROPOSITION 20.

1. 'In any triangle two sides taken together in any manner are greater than the remaining side.' The phraseology is similar to that of the enunciation of Prop. 17: see the note on that proposition. The meaning is that the sum of any two sides is greater than the third side.

According to Proclus, it was the habit of the Epicureans to ridicule this theorem as being evident even to an ass and requiring no proof, the argument being that, if fodder is placed at one angular point of a triangle and the ass is at another, he does not, in order to get to his food, traverse the two sides of the triangle but only the one side: an argument

which made Savile retort that its authors deserved
to be put on the same diet as the ass.

PROPOSITION 21.

1. ἀπὸ τῶν περάτων, 'from its extremities.'

2. ἐντὸς συσταθῶσιν. As we have seen (note on
Prop. 7), συνίστασθαι, 'to be constructed,' is
specially used of 'constructing' two straight lines
from the extremities of a straight line (in this case
a side of the triangle) to meet at a point. ἐντὸς
means 'within the triangle.'

We may translate 'If on one of the sides of a
triangle, from its extremities, there be constructed
two straight lines meeting within the triangle.'

2, 3. αἱ συσταθεῖσαι, 'the straight lines (so) constructed.'

3. τῶν λοιπῶν τοῦ τριγώνου δύο πλευρῶν, genitive
after ἐλάττονες, 'less than the remaining two sides
of the triangle.'

20. μείζονες ἐδείχθησαν, 'were proved (to be)
greater.'

25. διὰ ταὐτὰ τοίνυν, 'For the same reason then.'

PROPOSITION 22.

1. αἵ εἰσιν ἴσαι τρισὶ ταῖς δοθείσαις. We may
translate 'which are equal to three given straight
lines,' although the force of the words is rather
'which are equal to three straight lines, namely those
given' or 'to given straight lines three in number.'

3. δεῖ δὴ, 'Thus it is necessary that two of the
straight lines taken together in any manner should

be greater than the remaining (straight line).' What is introduced by the words δεῖ δὴ is the statement of the condition which is necessary in order that a solution of the problem may be possible. This, as we have seen (p. 161–2 above), was called the διορισμός. The criterion of possibility is of course supplied by the theorem in Prop. 20. If any two of the given straight lines are not greater than the third, no triangle can be constructed with sides equal to them respectively.

The MSS. add the following words to the enunciation: διὰ τὸ καὶ παντὸς τριγώνου τὰς δύο πλευρὰς τῆς λοιπῆς μείζονας εἶναι πάντῃ μεταλαμβανομένας. These words (διὰ τὸ... 'because of the fact that'... followed by the acc. and inf.) give a mere repetition of the enunciation of Prop. 20 and are probably a gloss; they are not found in Proclus.

12. ἐκκείσθω, for the passive of ἐκτιθέναι, as usual: 'let a straight line be set out.'

πεπερασμένη μὲν κατὰ τὸ Δ ἄπειρος δὲ κατὰ τὸ E, 'terminated at D but unlimited (in length) at (i.e. in the direction of) E.'

We have again, as in 1, 1, to satisfy ourselves that the two circles drawn as in the figure will actually meet in some point. They will in fact meet in two points, one on each side of DE, if the condition is fulfilled that any two of the straight lines A, B, C are together greater than the third. Euclid assumes this without proof.

PROPOSITION 23.

1. πρὸς τῇ δοθείσῃ εὐθείᾳ. In this proposition, where an angle has to be constructed *on* a straight line, i.e. so that the given straight line is one of the straight lines containing it, it is best to translate 'On a given straight line and at (πρὸς understood) a point on it.' The order of the rest of the enunciation is συστήσασθαι γωνίαν εὐθύγραμμον ἴσην τῇ δοθείσῃ γωνίᾳ εὐθυγράμμῳ, 'to construct a rectilineal angle equal to a given rectilineal angle.'

Observe that we have to construct the triangle so that *AG* lying along *AB* and equal to *CE* is the base, and *AF* is equal to *CD* and *GF* to *DE*; that is to say, we have to describe two circles (1) with *A* as centre and radius equal to *CD* and (2) with *G* as centre and radius equal to *DE*.

According to Proclus, Eudemus said that this problem was 'rather the discovery of Oenopides,' from which we may probably infer, not that the problem was the invention of Oenopides, but that he was the first to propound the solution here given by Euclid.

Apollonius, we are told, gave a different solution. Although, according to Proclus, the language used by Apollonius was inappropriate to Book I, the solution amounts to this. Let *CDE* be the given angle, *AB* the given straight line.

With *A* as centre and *AB* as radius describe a circle *BF*. Cut off *DC* equal to *AB* and with *D* as centre and *DC* as radius describe a circle meeting *DE* in *E*.

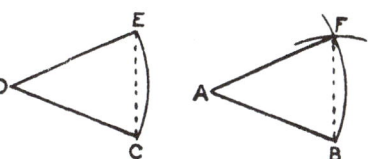

Join *CE*, and with *B* as centre and radius equal to *CE* draw a circle cutting the circle *BF* in *F*. Join *AF*.

Then the angle *BAF* is equal to the angle *CDE*; for in the triangles *DCE*, *ABF* two sides *CD*, *DE* are equal to two sides *BA*, *AF*, and the base *CE* is equal to the base *BF*, so that I, 8 applies.

This method was evidently intended to be more *practical* than Euclid's.

PROPOSITION 24.

The wording of the enunciation follows closely that of the enunciation of I, 4, with μείζονα, instead of ἴσην, qualifying the included angle and the base in one of the triangles. We may translate 'If two triangles have two sides equal to two sides respectively but have the one of the angles contained by the equal straight lines greater than the other, they will also have the base greater than the base.'

15. κείσθω ὁποτέρᾳ τῶν ΑΓ, ΔΖ ἴση ἡ ΔΗ, 'let *DG* be made equal to either of the (sides) *AC*, *DF*'; since *AC*, *DF* are equal, it does not matter whether *DG* is made equal to *AC* or to *DF*.

As the proposition is framed, there are three possible cases, of which, as usual, Euclid gives only one; the others are left for the student to supply for himself, and they are not difficult. In the case taken by Euclid *EG* falls within the angle *DEF*. The other two possibilities are these:

(1) *G* may fall on *EF* produced. In this case it is at once obvious that *EG* is greater than *EF*.

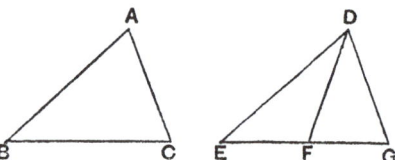

(2) *EG* may fall below *EF*, making the angle *DEG* greater than the angle *DEF*.

In this case, by I, 21, *DF*, *FE* are together less

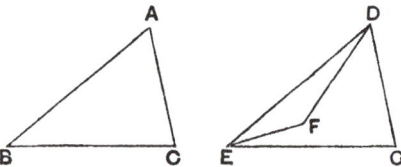

than *DG*, *GE*. But *DF* is equal to *DG*; therefore, subtracting the equals, we have *EF* less than *EG* or *BC*.

Simson, as is well known, reduced the number of cases to one by inserting, at the beginning, the

words 'Of the two sides *DE*, *DF* let *DE* be the side
which is not greater than the other.'

A modern proof covering both Euclid's case and
the case numbered (2) above is worth giving.

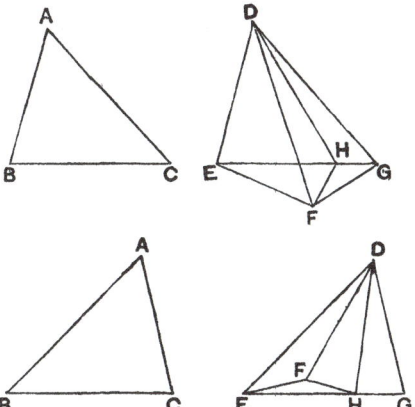

Make the triangle *DEG* equal in all respects to the
triangle *ABC*, as in the proposition.

Bisect the angle *FDG* by the straight line *DH*
meeting *EG* in *H*.

Then, in the triangles *FDH*, *GDH*,
the two sides *FD*, *DH* are equal to the two sides
GD, *DH*,
and the included angles *FDH*, *GDH* are equal.

Therefore the bases *HF*, *HG* are equal.   [1, 4]

Therefore *EG* is equal to the sum of *EH*, *HF*.

But *EH*, *HF* are together greater than *EF*; [1, 20]
therefore *EG*, or *BC*, is greater than *EF*.

PROPOSITION 25.

3. καὶ τὴν γωνίαν τῆς γωνίας μείζονα ἕξει τὴν ὑπὸ τῶν ἴσων εὐθειῶν περιεχομένην. We may translate 'they will also have the one of the angles contained by the equal straight lines greater than the other.'

This proposition is a partial converse of the preceding, and Euclid proves it by *reductio ad absurdum*. Two direct proofs are handed down in Proclus's Commentary. They are due to Menelaus and Heron (both of Alexandria) respectively.

PROPOSITION 26.

This is the third theorem proving that two triangles which have certain parts respectively equal are *congruent*, that is, equal in all respects. I, 4 proves that this is the case if two sides and the included angle are respectively equal; I, 8 shows that the two triangles are congruent if all three sides are equal respectively. In the present proposition the hypothesis is that the triangles have *two* angles and *one* side respectively equal, and there are two cases according to the particular sides which are assumed to be equal. The equal side is ἤτοι τὴν πρὸς ταῖς ἴσαις γωνίαις ἢ τὴν ὑποτείνουσαν ὑπὸ μίαν τῶν ἴσων γωνιῶν, 'either (1) the side adjoining (or adjacent to, πρὸς) the equal angles, or (2) the side subtending (i.e. opposite to) one of the equal angles' in each triangle.

29. ὑπόκειται ἴση, 'is by hypothesis equal' (literally 'is supposed equal'), ὑποκεῖσθαι being used as the passive of ὑποτιθέναι (cf. 'hypothesis').

The proofs are by *reductio ad absurdum* and, though somewhat long, do not require particular elucidation.

Another method in use in ancient times was based on the application of one triangle to the other: this is the method of superposition which there is reason to believe that Euclid avoided wherever he could.

Proclus has the following interesting note. 'Eudemus in his geometrical history refers this theorem to Thales. For he says that in the method by which they relate that Thales proved the distance of ships on the sea it was necessary to make use of this theorem.' This may have been done in either of two ways.

(1) The distance of the ship from the shore may

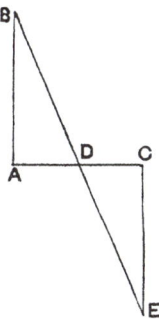

have been found thus. Let *B* be the ship, *A* a point on the shore. Take any distance *AC* on the shore at right angles to *AB*, and bisect *AC* at *D*. From *C* draw *CE* at right angles to *CA* in the direction away from B, and let E be the point on CE which is found to be in a straight line with B and D.

Then in the two triangles BAD, ECD two angles BAD, ADB are equal to two angles ECD, CDE respectively, and one side AD is equal to one side CD;

Therefore, by 1, 26, the triangles are equal in all respects, and CE (which can be measured) is equal to AB, the required distance.

(This was the method used by Marcus Junius Nipsus, a Roman agrimensor, to find the distance from a point A to an inaccessible point B.)

As however this method would require a certain extent of free and level ground to admit of the construction and measurements described, I suggest that the method of Thales may rather have been the following.

(2) If, as we may assume, he could reach an elevation, e.g. the top of a tower, he had only to use a rough instrument made of a straight stick with a straight cross-piece fastened to it so as to be capable of turning about the fastening (say a nail) so that it could form any angle with the stick and would remain where it was put. (Even a large pair of compasses would serve the same purpose.) The first thing then was to fix the stick upright (by means of a plumb-line) and direct the cross-piece to the ship. Next, leaving the cross-piece at the angle so found, the observer would turn the stick round, while keeping it always vertical, until the cross-piece pointed to some visible and accessible object on the

shore, when the object could be mentally noted and
the distance from the bottom of the tower to it
could be measured at leisure. This would, by 1, 26,
give the distance from the bottom of the tower to
the ship.

Professor David Eugene Smith has confirmed the
probability of this suggestion. He observes (*The
Teaching of Geometry*, 1911, p. 172) that this method
is found in so many practical geometries of the first
century of printing that it seems to have long been
a common expedient. He adds a story that one of
Napoleon's engineers won the imperial favour by
quickly measuring the width of a stream that
blocked the progress of the army, using this very
method.

PROPOSITION 27.

1. εἰς δύο εὐθείας ἐμπίπτουσα, 'falling on two
straight lines.' The phrase is the same as that used
in Post. 5, and means that the straight line *lies
across* the two straight lines, is, in fact, what we call
a *transversal*.

τὰς ἐναλλὰξ γωνίας, 'the alternate angles.' ἐναλλὰξ,
an adverb, 'alternately,' is here used adjectivally.
The same word is used of the transformation of a
proportion in four terms by which the terms are
taken *alternately*, i.e. if the proportion A is to B as
C is to D is altered to the proportion A is to C as B
is to D. The use of 'alternate' to describe the angles
AEF, EFD (or the angles BEF, EFC) in the figure

of the proposition is natural if we consider the four internal angles at E and F and take them (say) in the order of the motion of the hands of a watch, i.e. in the order AEF, FEB, EFD, EFC, when it is seen that AEF and EFD are alternate.

8. συμπεσοῦνται (future of συμπίπτω), 'will meet.'

9. ἐπὶ τὰ Β, Δ μέρη, 'in the direction of B, D' (literally, 'towards the parts, or regions, B, D') We must understand μέρη after 'Α, Γ' also: ἤτοι... ἢ..., 'in the direction either of B, D or of A, C.'

11. ἡ ἐκτὸς γωνία, 'the exterior angle'; ἡ ἐντὸς καὶ ἀπεναντίον γωνία, 'the interior and opposite angle,' as usual (cf. Prop. 16).

15, 16. ἐπὶ μηδέτερα τὰ μέρη, 'in neither direction,' as in Post. 5.

This proposition begins the second section of Book I. The first section has dealt mainly with triangles, their construction and their properties in the sense of the relation of their parts, the sides and angles, to one another, and the comparison of different triangles with reference to their parts and to their areas in the particular cases where they are congruent.

This second section establishes the theory of parallels and introduces the cognate matter of the equality of the sum of the three angles of any triangle to two right angles; then come two propositions, namely 1, 33, 34, which introduce the parallelogram for the first time and form the tran-

sition to the third and final section. The third
section treats of the areas of triangles, parallelograms and squares in relation to one another, the
special feature of the section being a new conception
of *equality* of areas, equality not dependent on
congruence.

PROPOSITION 28.

One criterion of the parallelism of two straight
lines, the equality of the alternate angles made
with any transversal, has been given in 1, 27; this
proposition gives two alternative tests which are
equivalent to that of 1, 27 and are in fact reduced
to it.

2. τῇ ἐντὸς καὶ ἀπεναντίον καὶ ἐπὶ τὰ αὐτὰ μέρη,
'(equal) to the interior and opposite (angle) *on the
same side*,' i.e. on the same side of the line (the
transversal) which 'falls on' (i.e. across) the two
others.

3. τὰς ἐντὸς καὶ ἐπὶ τὰ αὐτὰ μέρη, 'the interior
angles on the same side.'

From remarks on the criteria of parallelism made
by Aristotle (*Anal. Post.* I, 5 and *Anal. Prior.* II, 17)
it is clear that the two propositions 1, 27, 28 were
familiar to him; but, as we gather that the theory of
parallels current in his time contained some vicious
circle (an assumption, presumably in a different
form, of the very thing required to be proved), it
seems clear that Euclid's Postulate 5 and his proposition 1, 29 based upon it were due to Euclid

himself, who may therefore be taken to have placed the theory on a proper basis for the first time.

PROPOSITION 29.

This important proposition is the converse of 1, 27, 28 and depends for its proof on the famous Postulate 5. The three alternative criteria of parallelism are shown to be satisfied by any two parallel straight lines.

18. αἱ δὲ ἀπ᾽ ἐλασσόνων ἢ δύο ὀρθῶν ἐκβαλλόμεναι εἰς ἄπειρον συμπίπτουσιν, literally 'straight lines produced without limit from angles less than two right angles meet.' This is a variation from the more explicit language of Post. 5 but means the same thing. A good deal is left to be understood, namely that the straight lines start from points at which they meet a transversal, and that they make with it interior angles on the same side the sum of which is less than two right angles.

21. διὰ τὸ παραλλήλους αὐτὰς ὑποκεῖσθαι, 'because they are by hypothesis parallel' (lit. 'are supposed parallel').

In modern text-books Euclid's Postulate 5 is often replaced by another, known as Playfair's Axiom. It was however no new discovery of Playfair's, since it is explicitly stated in Proclus's Commentary on Book 1. It is to the effect that 'Through a given point only one parallel can be drawn to a given straight line' or 'If two straight lines intersect one another, they cannot both be parallel to one and the same straight line.'

If this Axiom is used, the proof of I, 29 proceeds thus.

Given two parallel straight lines *AB*, *CD* and a

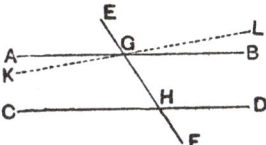

transversal *EF*, to prove that the alternate angles *AGH*, *GHD* are equal.

If they are not equal, draw another straight line *KL* through *G* making the angle *KGH* equal to the angle *GHD*.

Then, since the alternate angles *KGH*, *GHD* are equal,   *KL* is parallel to *CD*.   [1, 27]

Therefore two straight lines *KL*, *AB* intersecting at *G* are both parallel to *CD*: which is impossible (by the Axiom).

Therefore the angle *AGH* cannot but be equal to the angle *GHD*.

The rest of the proposition follows as in Euclid.

PROPOSITION 31.

This proposition implies, as Proclus observes, that only one straight line can be drawn through a given point parallel to a given straight line.

PROPOSITION 32.

1–3. 'In any triangle, if one of the sides be produced, the exterior angle is equal to the two interior and opposite angles,' i.e. to the *sum* of the two interior and opposite angles.

According to Proclus, Eudemus referred the discovery of this theorem to the Pythagoreans, and gave what he affirmed to be their demonstration of it (see below). It is a question whether Thales was aware of the theorem, so far at least as the right-angled triangle is concerned; for Pamphile (reign of Nero, A.D. 54–68) is quoted by Diogenes Laertius as saying that Thales, who learnt geometry from the Egyptians, was the first to inscribe a right-angled triangle in a circle and sacrificed an ox on the strength of it; in other words, that he proved that the angle in a semicircle is right (whence it would easily be seen that the sum of the three angles of a *right-angled* triangle is equal to two right angles). But other authorities said it was Pythagoras.

The Pythagorean proof of the equality of the sum of the three angles of any triangle to two right angles was, according to Eudemus, the following.

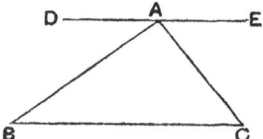

Let *ABC* be any triangle, and through *A* draw *DAE* parallel to *BC*.

Then, since *BC*, *DE* are parallel, the alternate angles *DAB*, *ABC* are equal to one another; and so are the alternate angles *EAC*, *ACB* also.

Therefore the sum of the angles *ABC*, *ACB* is equal to the sum of the angles *DAB*, *EAC*.

Add to each the angle *BAC*; therefore the sum of the angles *ABC*, *ACB*, *BAC* is equal to the sum of the angles *DAB*, *BAC*, *CAE*, that is, to two right angles.

We gather from a remark of Aristotle that the proof with which he was familiar followed Euclid's method.

PROPOSITION 33.

1. αἱ τὰς ἴσας τε καὶ παραλλήλους ἐπὶ τὰ αὐτὰ μέρη ἐπιζευγνύουσαι εὐθεῖαι, literally 'the straight lines joining equal and parallel (straight lines) in the same directions'; by this is meant, and we may so translate, 'the straight lines joining (those extremities of) equal and parallel straight lines (which are) in the same directions.'

2, 3. ἴσαι τε καὶ παράλληλοί εἰσιν, 'are (themselves) equal and parallel.'

5. ἐπιζευγνύτωσαν αὐτὰς ἐπὶ τὰ αὐτὰ μέρη εὐθεῖαι αἱ ΑΓ, ΒΔ, 'let the straight lines *AC*, *BD* join them (i.e. their extremities) in the same directions.' The meaning is that *A*, *C* and *B*, *D* are joined respec-

tively, not B, C and A, D, which pairs of points are not 'in the same directions.'

12. δύο δὴ αἱ ΑΒ, ΒΓ δύο ταῖς ΒΓ, ΓΔ ἴσαι εἰσίν, 'the two sides AB, BC are equal to the two sides BC, CD'; we should naturally say DC, CB instead of BC, CD in order to put the equals in corresponding order.

The effect of this proposition is to demonstrate the existence of the figures which we call parallelograms.

PROPOSITION 34.

The preceding proposition has shown a method of constructing a four-sided figure with its opposite sides parallel, and Euclid in this proposition uses the terms 'parallelogrammic area' (or, more exactly, 'parallelogram area,' παραλληλόγραμμον χωρίον, meaning an area formed by parallel lines, in pairs) and 'parallelogram' (παραλληλόγραμμον alone) to describe the said figure, without any definition or further explanation. The word is tacitly appropriated to *four-sided* figures with opposite sides parallel; and we gather from Proclus that Euclid was the first to introduce the word παραλληλό-γραμμον in this restricted sense, which excludes such figures with more than four sides, e.g. a regular hexagon, as have their opposite sides parallel.

1. χωρίον means any 'place' or 'area,' but is, in Greek geometry, usually restricted to areas in the shape of rectangles or parallelograms.

1. αἱ ἀπεναντίον πλευραί τε καὶ γωνίαι, 'the opposite sides and angles (respectively).'

3. διάμετρος, 'diameter.' We generally use the word 'diagonal' in relation to a parallelogram; the Greek usage preferred 'diameter,' and the word should be so translated here.

αὐτὰ δίχα τέμνει, 'bisects them' (i.e. the parallelogrammic areas or parallelograms).

17. μίαν πλευρὰν μιᾷ πλευρᾷ ἴσην τὴν πρὸς ταῖς ἴσαις γωνίαις κοινὴν αὐτῶν τὴν ΒΓ, 'one side equal to one side, namely that which adjoins the equal angles (and is) common to the triangles, that is, BC.'

PROPOSITION 35.

Euclid here drops the longer expression 'parallelogrammic area' and uses the word 'parallelogram' exclusively.

2. ἐν ταῖς αὐταῖς παραλλήλοις, 'in the same parallels,' means included by the same two parallel straight lines; that is, the side opposite the base always lies along one and the same straight line which is parallel to the base.

ἴσα ἀλλήλοις ἐστίν, 'are equal to one another.'

We are here introduced to a new sense of the word 'equal' as applied to two figures. Hitherto we have had equality in the sense of congruence or symmetry only; the equal triangles, as well as the equal straight lines and the equal angles, which we have met with up to now have all been equal in this sense. But now, without any explicit warning of a change in the

meaning of the word, figures are inferred to be
'equal' which are equal in area or content but need
not be of the same *shape*.

In this proposition the. 'equality,' in the new
sense, of the parallelograms *ABCD*, *EBCF* is inferred by means of two steps. First, one and the
same area, the small triangle *DGE*, is subtracted
from each of the two triangles *ABE*, *DCF* which are
equal in the sense of *congruent*, and it is inferred that

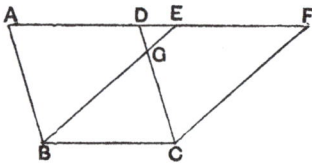

the remainders, the trapezia *ABGD*, *EGCF*, which
are not of the same shape or congruent, are 'equal'
in the new sense. Then, secondly, to each of these
'equals' is added one and the same area, the triangle
GBC, and it is inferred that the sums, which are the
two parallelograms respectively, are also 'equal' in
the new sense.

Simson (after Clairaut) slightly altered the proof
by substituting *one* step of subtracting congruent
areas (the triangles *AEB*, *DFC*) from one and the
same area (the large trapezium *ABCF*) for the *two*
steps in Euclid's proof.

While, in either case, nothing more is explicitly
used than the axioms that, 'if equals be added to

equals, the wholes are equal' and 'if equals be subtracted from equals, the remainders are equal,' there is the further *tacit* assumption that it does not matter to what part or from what part of the same or equal areas the same or equal areas are added or subtracted. De Morgan observed that the postulate 'an area taken from the same area leaves the same area from whatever part it may be taken' is particularly important as the key to equality of rectilineal areas which could not be cut into parts that are respectively coincident.

As Proclus points out, there are three cases of this proposition according to the form which the figure

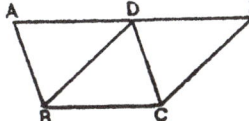 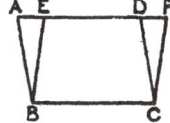

takes. Euclid, as usual, takes one case only and leaves the others, shown in the accompanying figures, for the student to investigate for himself.

Proclus observes that the present theorem and the similar proposition (37) regarding triangles are among the theorems in mathematics which were deemed paradoxical, because the uninstructed might well think it impossible that the area of the parallelograms should remain the same while the length of the sides other than the base and the side opposite to it may increase indefinitely. In fact the sum of the sides, or the *perimeter*, of parallelograms or triangles is of itself no criterion as to their area. Misconceptions on the subject were rife among non-mathematicians; Proclus tells us that people used to judge of the size of cities by their perimeters, and certain members of communistic societies in his own time cheated their fellow-members by giving them land of greater perimeter but less area than that which they took themselves, so that, on the one hand, they got a reputation for greater honesty, while, on the other, they took more than their share of the produce. Literature furnishes other examples of the same misconception; thus Thucydides estimated the size of Sicily by the time required to circumnavigate it.

PROPOSITION 37.

The equality of the triangles ABC, DBC is inferred from the facts (1) that the parallelograms $EBCA$, $DBCF$ are equal, and (2) that the triangle ABC is half of the former parallelogram, while the triangle DBC is half of the latter. The inference is, in the Greek text, justified by the remark (18) τὰ δὲ τῶν ἴσων ἡμίση ἴσα ἀλλήλοις ἐστίν, 'But the halves of equals are equal to one another.' This is a reference to an axiom which was apparently interpolated among Euclid's *Common Notions* in very early times. It is probable therefore that the reference to it here was also interpolated, and that Euclid himself made the inference without any explanation.

A similar remark applies to the final step in the next proposition (38).

PROPOSITION 39.

We translate the enunciation thus, 'Equal triangles which are on the same base and on the same side are also in the same parallels.'

After the words (5) ἐπὶ τὰ αὐτὰ μέρη τῆς ΒΓ in the *setting-out*, the text proceeds λέγω ὅτι καὶ ἐν ταῖς αὐταῖς παραλλήλοις ἐστίν. Ἐπεζεύχθω γὰρ ἡ ΑΔ· λέγω ὅτι παράλληλός ἐστιν ἡ ΑΔ τῇ ΒΓ, 'I say that they are also in the same parallels. For let *AD* be joined; I say that *AD* is parallel to *BC*.' The two statements of the fact to be proved, introduced by λέγω, 'I say,' are suspicious; and Heiberg has proved by means of an ancient papyrus-fragment that the words 'I say that they are also in the same parallels' are an interpolation by some one who did not observe that the words 'And let *AD* be joined' are part of the *setting-out* but took them to belong to the *construction*; the interpolator then altered 'And' into 'For.' We should therefore, after τῆς ΒΓ, read as follows: καὶ ἐπεζεύχθω ἡ ΑΔ· λέγω ὅτι παράλληλός ἐστιν ἡ ΑΔ τῇ ΒΓ, 'And let *AD* be joined. I say that *AD* is parallel to *BC*.'

The theorem is a partial converse of I, 37. In I, 37 we have triangles which are (1) on the same base, (2) in the same parallels, and the theorem proves (3) that the triangles are equal. The partial converse takes as hypotheses the hypothesis (1) and the conclusion (3), and proves (2) as the conclusion.

PROPOSITION 40.

Heiberg has shown by means of the papyrus-fragment mentioned in the last note that this proposition is an interpolation by some one who thought that there should be a proposition following 1, 39 and related to it in the same way as 1, 38 is related to 1, 37 and 1, 36 to 1, 35.

PROPOSITION 41.

1. ἐὰν παραλληλόγραμμον τριγώνῳ βάσιν τε ἔχῃ τὴν αὐτὴν καὶ ἐν ταῖς αὐταῖς παραλλήλοις ᾖ. The dative τριγώνῳ is constructed with τὴν αὐτὴν and ἐν ταῖς αὐταῖς παραλλήλοις. 'If a parallelogram have the same base, and be in the same parallels, *with* a triangle' (or '*as* a triangle').

Euclid infers that, because the parallelogram is double of the triangle ABC, and the triangle ABC is equal to the triangle EBC, the parallelogram is also double of the triangle EBC. This can be deduced by means of Common Notions 1 and 2, in this way. Triangle ABC = triangle EBC; therefore (adding equals to equals, C. N. 2) twice triangle ABC = twice triangle EBC. But parallelogram = twice triangle ABC; therefore parallelogram = twice triangle EBC (C. N. 1).

PROPOSITION 42.

1. The order of words in the enunciation is συστήσασθαι παραλληλόγραμμον ἴσον τῷ δοθέντι τριγώνῳ, 'to construct a parallelogram equal to a given triangle.'

2. ἐν τῇ δοθείσῃ γωνίᾳ εὐθυγράμμῳ, 'in a given rectilineal angle'; this means having one of its angles equal to a given rectilineal angle, so that the parallelogram could be conceived to be *placed in* the angle in the sense of having one of its angles coincident with the angle.

PROPOSITION 43.

1. It is best to translate παντὸς παραλληλο-γράμμου by 'In any parallelogram,' as usual.

τὰ παραπληρώματα τῶν περὶ τὴν διάμετρον παραλ-ληλογράμμων, 'the *complements* of the parallelo-grams about the diameter.' The complements are the parts which fill up the interstices (παρά and πλήρης, 'full'), i.e., when added to the 'parallelo-grams about the diameter,' they complete the original parallelogram.

6. In the *setting-out* the complements are de-scribed as 'the so-called complements,' τὰ λεγόμενα παραπληρώματα: 'let *BK, KD* be the so-called com-plements.' This sufficiently explains what is meant; the complements are the figures (parallelograms) left on each side of the diameter when the 'para-lelograms about the diameter,' *EH* and *GF*, are taken away from the original parallelogram.

This seemingly insignificant theorem is, in the next proposition, made the basis of one of the most important and far-reaching problems of construction in the whole of theoretical geometry.

PROPOSITION 44.

We are here introduced to an elementary case of the general method of *application of areas* which is fundamental in Greek geometry.

1, 2. παραβαλεῖν παρὰ, literally 'to lay along-side of' or, as we generally say, 'to apply to.' 'To apply to a given straight line a parallelogram equal to a given triangle in a given rectilineal angle.' The last phrase, 'in a given rectilineal angle,' means, as in Prop. 42, that the parallelogram must be such that one of its angles is equal to a given rectilineal angle.

15. ἐνέπεσεν: we may translate 'falls,' although Euclid uses the aorist, 'fell' or 'has fallen.'

17. αἱ δὲ ἀπὸ ἐλασσόνων ἢ δύο ὀρθῶν κ.τ.λ. Post. 5 is here cited, as in Prop. 29, in shorter but equivalent terms: 'straight lines produced without limit from angles less than two right angles meet.'

24. περὶ δὲ τὴν ΘΚ παραλληλόγραμμα μὲν τὰ ΑΗ, ΜΕ, τὰ δὲ λεγόμενα παραπληρώματα τὰ ΛΒ, ΒΖ. The words περὶ τὴν ΘΚ, 'about *HK*,' go with both the expressions distinguished by μὲν and δὲ: 'and *AG*, *ME* are parallelograms, and *LB*, *BF* the so-called complements, about *HK*.'

It is necessary to lay stress on the supreme importance of this proposition. Prop. 42 has shown us

how to make a parallelogram equal to a given triangle and having one angle equal to a given angle.
But the length of the sides of that parallelogram are
determined for us by the construction, since one of
them is *necessarily* equal to half one of the sides of
the given triangle, and the construction determines
the length of the other; we cannot, by means of that
proposition, make either side of the parallelogram
of any length we please. This is what the present
proposition enables us to do; whatever shape the
parallelogram in Prop. 42 has, we can now transform
it into another with the same angle and of equal
area but with *one side of any required length*, e.g. *a
unit* length, say an inch, a foot or a centimetre long;
and the combined effect of the two propositions is
that we can construct a parallelogram having one
side of any length we please and one angle equal to
any angle we please, which shall be equal in area to
any given triangle. The marvellous ingenuity of the
solution of the problem by the simple application of
the property that the 'complements of the parallelograms about the diameter' are equal is indeed
worthy of the 'godlike men of old,' as Proclus calls
the discoverers of the method of *application of areas*.

We have the authority of Eudemus for the statement that the discovery of the method of *application of areas* was due to the Pythagoreans, and there
would seem to be no reason to doubt that the particular solution of the problem in 1, 44 is that given
by the Pythagoreans.

The present problem is however only the most
elementary case of the complete method of *application of areas* discovered by the Pythagoreans, which,
taken as a whole, is one of the most effective methods
employed by the Greeks in higher as well as elementary geometry. Not only did they *apply* parallelograms of given area to a given straight line exactly,
but they applied parallelograms equal to any given
rectilineal figure in such a way that they overlapped
or fell short of the straight line to which they were
applied, and that the portion of the parallelogram
which overlapped or fell short had its sides in a given
ratio. The extended method requires propositions
in Book II and Book VI of Euclid and therefore cannot be described here. It must suffice to say that the
problems solved by the method are the geometrical
equivalent of the solution of mixed quadratic
equations in algebra.

Apollonius used the method as the foundation of
his treatment of conic sections. The overlapping of
the area was called *exceeding*, ὑπερβολή, the *falling-
short* was called ἔλλειψις, and the exact *application*
was παραβολή; and these terms (in English *hyperbola, ellipse* and *parabola*) were actually used for the
first time by Apollonius to describe the three different
conic sections.

PROPOSITION 45.

Any rectilineal figure can be divided up into a
certain number of triangles; and this proposition

shows how, by means of I, 42 and 44, we can make *one* parallelogram with one side of given length and one angle equal to a given angle which shall be equal to the whole rectilineal figure.

2, 3. εὐθύγραμμον, literally 'rectilineal,' is used in the enunciation and later on as a substantive to mean a 'rectilineal figure,' just as παραλληλό-γραμμον is used for a parallelogram.

PROPOSITION 46.

1. ἀναγράφειν ἀπό, literally 'to draw up *from*,' is the technical expression for ' describe upon.' ' On a given straight line to describe a square.' The expression is to be distinguished from συστήσασθαι, which is used of *constructing* a triangle (or an angle). The triangle is, so to speak, *pieced together*, while the describing of a square on a given straight line is the making of a figure 'from' *one* side.

As before remarked, the definition of a square (Def. 22) says nothing as to whether a figure corresponding to the definition exists or not. The proof that such a thing *exists* is furnished by the actual construction of it in this proposition; for the figure drawn is proved to be both equilateral and right-angled (in the sense of having all its angles right angles).

PROPOSITION 47.

1. τὸ ἀπὸ τῆς τὴν ὀρθὴν γωνίαν ὑποτεινούσης πλευρᾶς τετράγωνον, 'the square on the side subtending the right angle.' As 'to describe a square

on' a straight line is ἀναγράφειν τετράγωνον ἀπό, so τὸ ἀπὸ τῆς εὐθείας ἀναγραφὲν or ἀναγεγραμμένον τετράγωνον is the full expression for 'the square described on the straight line.' The participle is generally omitted, as here; and τετράγωνον itself is commonly left to be understood, τὸ ἀπὸ (with gen.) meaning invariably 'the square on.'

ἡ τὴν ὀρθὴν γωνίαν ὑποτείνουσα πλευρά is 'the side subtending the right angle,' ὑποτείνουσα being here used with the simple acc. instead of with ὑπὸ and acc. ὑποτείνουσα in connexion with a right-angled triangle came to be used alone in the special sense of *hypotenuse*, meaning the particular side which 'subtends,' or is opposite to, the right angle.

3. τοῖς ἀπὸ τῶν τὴν ὀρθὴν γωνίαν περιεχουσῶν πλευρῶν τετραγώνοις, 'to the squares on the sides containing the right angle.'

The full enunciation then is 'In right-angled triangles the square on the side subtending the right angle is equal to the squares on the sides containing the right angle.'

11. ὁποτέρᾳ τῶν ΒΔ, ΓΕ, 'to either of the straight lines *BD*, *CE*.'

16, 17. ἐπ' εὐθείας ἄρα ἐστὶν ἡ ΓΑ τῇ ΑΗ, 'Therefore *CA* is in a straight line with *AG*.'

22. δύο δὴ αἱ ΔΒ, ΒΑ δύο ταῖς ΖΒ, ΒΓ ἴσαι εἰσὶν ἑκατέρα ἑκατέρᾳ, 'the two sides *DB*, *BA* are equal to the two sides *FB*, *BC* respectively.' The sides are not given in corresponding order; we should rather say 'the two sides *AB*, *BD*....'

33. τὰ δὲ τῶν ἴσων διπλάσια ἴσα ἀλλήλοις ἐστίν,
'But the doubles of equals are equal to one another.'
These words are no doubt interpolated; the in-
ference easily follows from Common Notion 2.

This great theorem is universally associated with
the name of Pythagoras. Proclus says 'If we listen
to those who wish to recount ancient history, we
find some of them referring this theorem to Pythagoras and saying that he sacrificed an ox in cele-
bration of his discovery.'   Plutarch, Diogenes
Laertius and Athenaeus also agree in attributing
the proposition to Pythagoras. These are, it is true,
somewhat late witnesses, but Cicero (*De natura
deorum*), while he disputes the story of the sacrifice
(which would have been inconsistent with Pythagorean ritual), does not seem to question the fact of
the geometrical discovery. Cicero is apparently
commenting on the verses of Apollodorus 'the cal-
culator' (ὁ λογιστικός), which allude to a discovery
of Pythagoras in the words 'when Pythagoras discovered the far-famed figure, that on account of
which he made the famous sacrifice of an ox.'
Plutarch, who quotes the same distich, is not certain
whether the sacrifice related to the discovery of the
proposition about the square on the hypotenuse or
to the solution of the problem about the 'application of an area' (by which he means the problem
'given two rectilineal figures, to describe a third
which shall be equal in area to one of the two given
figures and similar to the other'). But Plutarch

says nothing to suggest that he had any hesitation in attributing *both* propositions to Pythagoras.

The connexion of Pythagoras with 1, 47 is further confirmed by the tradition crediting him with a rule for finding a series of square numbers which are the sums of two squares, in other words, for finding any number of right-angled triangles with all their sides expressible in rational numbers. This theory is connected with the use of the *gnomon* in its geometrical and arithmetical sense. According to Aristotle the *gnomon* is the figure which, when placed round two sides of a square makes a larger square, while in arithmetic the *odd numbers* are gnomons. This is easily seen from the accompanying figures. Three

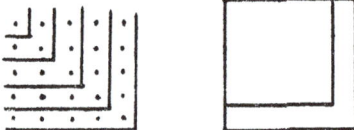

dots placed round one make 4 ($= 2^2$); five more placed round the 4 make 9, or 3^2; seven more round the 9 make 16, or 4^2; and so on. The geometrical gnomon shown in the second figure is regularly employed in Euclid Book II, and we know that it was a term well understood by the Pythagoreans.

The rule attributed to Pythagoras for finding right-angled triangles in rational numbers connects itself naturally with the use of the successive odd numbers as gnomons.

Since $\quad 1 = 1^2,$

$\quad\quad\quad 1 + 3 = 2^2,$

$\quad\quad\quad 1 + 3 + 5 = 2^2 + 5 = 3^2,$

$\quad\quad\quad 1 + 3 + 5 + 7 = 3^2 + 7 = 4^2,$

and so on, we easily deduce the general formula

$$a^2 + (2a + 1) = (a + 1)^2.$$

If this formula is to give a sum of two squares equal to a square, we have simply to make $(2a + 1)$ a square.

We write therefore

$$2a + 1 = n^2,$$

whence $\quad\quad a = \tfrac{1}{2}(n^2 - 1);$

therefore $\quad a + 1 = \tfrac{1}{2}(n^2 + 1),$

and the above formula becomes

$$[\tfrac{1}{2}(n^2 - 1)]^2 + n^2 = [\tfrac{1}{2}(n^2 + 1)]^2, \quad (n\text{ odd})$$

which is the precise formula attributed to Pythagoras.

In view therefore of the unanimity of tradition on the subject, there seems to be no sufficient reason to question that, so far as Greek geometry is concerned, Pythagoras was the first to introduce the theorem of 1, 47 and to give a general proof of it.

Not that particular cases of the theorem were not known long before Pythagoras. The triangle 3, 4, 5 was apparently recognised as right-angled in Egypt as far back as 2000 B.C. The ancient Babylonians were also probably aware of this case; the Chinese certainly knew it. The Indians too, long before the date of Pythagoras, had independently investigated, at least empirically, many cases of right-angled triangles. They even enunciated the theorem in all its generality, although there is no trace of their having given a scientific proof of it.

It is not possible to say with certainty how Pythagoras came to discover the general proposition or how he proved it. It is certain that the proof in Eucl. 1, 47 is Euclid's own. A proof identical *in substance* with Euclid's can be derived from propositions in Book VI (Props. 8, 4 and 17), and it is likely enough that the original proof used the method of proportions, and that Euclid ingeniously transformed this proof into that in 1, 47 which uses the methods of Book 1 only.

A most interesting and withal easy extension of this proposition to the case of parallelograms drawn on the sides of any triangle (not necessarily right-angled) is given by Pappus.

Pappus enunciates the theorem as follows.

If *ABC* be a triangle, and any parallelograms whatever *ABED*, *BCFG* be described on *AB*, *BC*, and if *DE*, *FG* be produced to meet at *H*, and *HB* be joined, the parallelograms *ABED*, *BCFG* are equal to the parallelogram contained by *AC*, *HB* in an angle which is equal to the sum of the angles *BAC*, *DHB*.

Produce *HB* to *K*; through *A*, *C* draw *AL*, *CM* parallel to *HK*, and join *LM*, meeting *HK* in *N*.

Then, since *ALHB* is a parallelogram, *AL*, *HB* are equal and parallel.

Similarly MC, HB are equal and parallel.
Therefore AL, MC are equal and parallel;

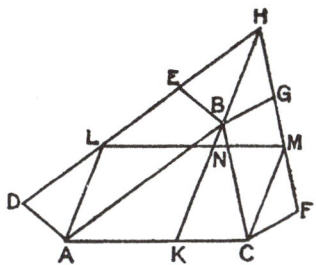

whence LM, AC are also equal and parallel,
and $ALMC$ is a parallelogram.

Further, the angle LAC of this parallelogram is
equal to the sum of the angles BAC, DHB, since
the angle DHB is equal to the angle LAB.

Now, since the parallelogram $DABE$ is equal to
the parallelogram $LABH$ (for they are on the same
base AB and in the same parallels AB, DH),
and likewise $LABH$ is equal to $LAKN$ (for they
are on the same base LA and in the same parallels
LA, HK),
the parallelogram $DABE$ is equal to the parallelogram $LAKN$.

For the same reason,
the parallelogram $BGFC$ is equal to the parallelogram $NKCM$.

Therefore the sum of the parallelograms $DABE$,
$BGFC$ is equal to the parallelogram $LACM$, that is,
to the parallelogram which is contained by AC, HB

in an angle *LAC* which is equal to the sum of the
angles *BAC*, *BHD*.

'And this,' says Pappus, 'is far more general than
what is proved in the Elements about squares in the
case of right-angled triangles.'

The accompanying figure shows the adaptation of
Pappus's proof to the particular case of I, 47.

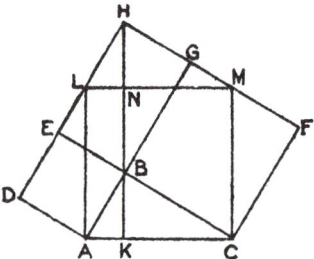

BD, *BF* are the squares on *AB*, *BC*, and we have
first to prove that, if *AL*, *CM* are drawn parallel to
BH to meet *DE*, *FG* in *L*, *M*, then *ALMC* is a
square.

By the construction, *BEHG* is a parallelogram, so
that $EH = BG = BC$.

And $BE = AB$.

Therefore in the two triangles *BEH*, *ABC*, two
sides *BE*, *EH* are equal to two sides *AB*, *BC*, while
the included angles are equal.

Therefore the triangles are equal in all respects,
so that the angle *EHB* is equal to the angle *ACB*.
(And $BH = AC$.)

Also, by construction, $ALHB$ is a parallelogram, so that the angle EHB is equal to the angle LAB.

Therefore the angle LAB is equal to the angle ACB.

Add to each the angle BAC;
therefore the angle LAC is equal to the sum of the angles BAC, ACB, and is therefore a right angle.

And $ALMC$ is a parallelogram, as in Pappus's proposition, while AL and CM are each equal to BH, which is equal to AC.

Therefore $ALMC$ is the square on AC.

We have now

$$\text{square } BD = \text{parallelogram } ALHB$$
$$= \text{parallelogram } ALNK,$$

and $\text{square } BF = \text{parallelogram } CMHB$
$$= \text{parallelogram } CMNK;$$

and, by addition, the sum of the squares BD, BF is equal to $ALMC$, which is the square on AC.

The best known alternative proof of I, 47 is that which is attributed to Thābit b. Qurra (826–901 A.D.). This proof has the effect of showing the equivalence of the sum of the two squares to the one square in a form which appeals directly to the eye.

Thābit proceeds in this way.

Let ABC be the given triangle, right-angled at A.

Produce AC to F, making EF equal to AC.

Describe on AB the square $ABDE$, and on EF the square $EFGH$.

Join *GC*. Produce *EH* (which clearly lies along *ED*) to *K*, so that *DK* may be equal to *AC*.

Then, since *KD* = *EH*, *HK* = *ED* or *AB*.

Therefore, in the four triangles *BAC*, *CFG*, *KHG*, *BDK*,

the sides *BA*, *CF*, *KH*, *BD* are all equal,

and the sides *AC*, *FG*, *HG*, *DK* are all equal.

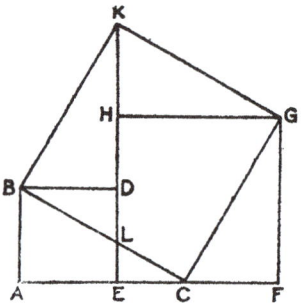

The equal sides also include equal angles (namely right angles) in all the triangles.

Therefore the four triangles are equal in all respects.

Hence *BC*, *CG*, *GK*, *KB* are all equal.

Again the angles *KBD*, *CBA* are equal;
add to each the angle *DBC*;
therefore the angle *KBC* is equal to the angle *ABD*, and is therefore a right angle.

Similarly the angle *CGK* is right.

Consequently *BCGK* is a square.

Now, taking the two squares *FH*, *EB* alongside

one another, and subtracting the two triangles *FCG*, *ABC*, we have as the remainder the quadrilateral *GCLH* and the triangle *LBD*. Adding to this remainder the two equal triangles *DBK*, *HKG* (these triangles being also equal to the triangles *FCG*, *ABC* which were before subtracted), we have as the sum the square *BG*.

It follows that the latter square, the square on *BC*, is equal to the sum of the two other squares *FH*, *EB*, which are the squares on *AC*, *AB* respectively.

To show the equivalence, we have only to take the squares *FH*, *EB* as placed, cut off the two triangles *CFG*, *BAC* and replace them in the positions *KHG*, *BDK*.

Although this proof is first found in the Arabic, it may well have been derived from a Greek source.

PROPOSITION 48.

This theorem is the converse of 1, 47 and presents no difficulty. It is however worthy of remark that, by drawing *AD* at right angles to *AC* on the side of it opposite to *AB*, Euclid avoids the appearance of indirect demonstration, because he has only to prove that the two right-angled triangles are equal in all respects. This he does by means of 1, 8. If, on the other hand, he had drawn *AD* on the same side as *AB*, it would have been necessary to prove that the two triangles coincide, and this would have been done by *reductio ad absurdum*. For, if the two triangles did not coincide, we should have, on the same side of CA, two pairs of straight lines CB, AB and CD, AD such that $CB = CD$ and $AB = AD$, which is impossible by I, 7 (the very proposition on which I, 8 depends).

INDICES

INDEX OF GREEK TERMS
EXPLAINED IN THE NOTES

Ἄγειν, to draw. ἠγμένη, 132; ἀγαγεῖν, 145
ἀδύνατος, -ον, impossible. ἡ εἰς τὸ ἀδύνατον ἀπαγωγή, reduction to the impossible, i.e. *reductio ad absurdum*, 172
αἰτεῖν, to demand or postulate, 142; ᾐτήσθω, 142
αἴτημα, a postulate, 142
ἄλλος, other. πρὸς ἄλλῳ καὶ ἄλλῳ σημείῳ, to different points, 174
ἀμβλυγώνιος, -ον, obtuse-angled, 137
ἀμβλύς, -εῖα, -ύ, obtuse, 127
ἀναγράφειν ἀπό, to describe on, 217
ἄνισος, unequal, 165
ἀντιστροφή, conversion, 172
ἀντίστροφος, -ον (adj.), converse, 171
ἀπαγωγή, reduction (of one problem to another, or of a hypothesis to an absurdity, *reductio ad absurdum*), 172
ἄπειρος, -ον, unlimited, 140, 182, 191; ἐπ᾽ ἄπειρον, εἰς ἄπειρον, *ad infinitum*, 140, 150
ἀπεναντίον (adv. used adjectivally), opposite, 139, 187, 207
ἀπλατής, -ές, breadthless, 115
ἀπό (c. gen.), on (constructed with ἀναγράφειν), 217; τὸ ἀπὸ τῆς ΒΓ (τετράγωνον) = the square on BC
ἀπόδειξις, *proof*, 161
ἀπολαμβάνειν, to cut off, 134
ἅπτεσθαι, to meet (of lines), 124; dist. ἐφάπτεσθαι
ἄρα, therefore, 159
ἀρχή, beginning. αἱ ἐξ ἀρχῆς εὐθεῖαι, the original straight lines, 174
ἀφαιρεῖν, to take away or subtract. ἀφαιρεθῇ, 152; ἀφῃρήσθω, 184

Βάσις, base, 166; base of an *angle*, 169

ἐπί (c. acc.), *to* (in 'perpendicular to'=let fall *on*), 127, 182; (in 'joined to'), 159; *towards* (see μέρη) (c. acc.) *on* (in 'set up on'), 127, 184
   (c. gen.) *on* (in 'construct on'), 157, 173, (= situated on), 182; *in* (in 'in a straight line,' ἐπ' εὐθείας), 146, 163, 218
ἐπιζευγνύναι, to join, 205; ἐπεζεύχθω, -ωσαν (ἐπί and acc.), 159, 211
ἐπίπεδος, -ον, plane, 123; ἐπ. ἐπιφάνεια, 123; γωνία, 124; σχῆμα, 129
ἐπιφάνεια, surface, 121. ἐπίπεδος ἐπ., 123
ἑτερόμηκες, oblong, 138, 139
εὐθεῖα (fem. of εὐθύς), sc. γραμμή, straight line, 118–121; ἐπ' εὐθείας, 'in a straight line,' 146, 163, 218
εὐθύγραμμος, -ον, rectilineal, 126; with γωνία, 126; σχῆμα, 135; alone (in neuter), 217
ἐφάπτεσθαι, to touch, 124
ἐφαρμόζειν (ἐπί and acc.), to coincide (with); ἐφαρμόζεσθαι, to be applied to, 153–4, 167–8, 176
ἐφεξῆς, successively, 126; adv. used as adj., αἱ ἐφ. γωνίαι, the adjacent angles, 126, 181, 185
ἐφιστάναι, to set up. ἐφεστηκυῖα, standing on (of a straight line), 126

Ἡμικύκλιον, semicircle, 134

Θεώρημα, theorem, 156

Ἰσόπλευρος, -ον, equilateral, 136
ἴσος, equal. ἐξ ἴσου, evenly, 118–19
ἰσοσκελής, -ες, isosceles, 136
ἱστάναι, to set up. σταθεῖσα, 126, 184

Κάθετος, ἡ, perpendicular, 127, 183
καθιέναι, to let fall, 126, 183
καί, then, 159
κατά (c. acc.), according to. κατὰ γνώμονα, gnomonwise, 127, 183; κατὰ κορυφήν, vertically opposite (of angles), 185; καθ' ὅ (σημεῖον), the point in (or *at*) which, 158
καταλείπειν, to leave behind. τὰ καταλειπόμενα, 152

Τετράγωνον (neut. of τετράγωνος), a square, 136, 139
τετράπλευρος, -ον, four-sided. τετράπλευρον (σχῆμα), quadrilateral, 135–6, 139
τιθέναι, to place. θέσθαι (mid.), 162
τομή, (point of) section, 186
τραπέζιον, trapezium, 138
τρίγωνος, -ον, triangular. τρίγωνον, triangle, 136
τρίπλευρος, -ον, three-sided, 135
τυγχάνω, to happen or chance. τυχὸν σημεῖον, a point (taken) at random, 169, 180

Ὑπερβολή, exceeding (in application of areas); (in conics, a hyperbola), 216
ὑπό (c. acc.), under or contained by. ἡ ὑπὸ ΒΑΓ γωνία, the angle BAC, 166, 167
ὑποκεῖσθαι, to be supposed or assumed (=pass. of ὑποτιθέναι). ὑπόκειται, is by hypothesis, 196, 202
ὑποτείνειν, to be opposite to or subtend. c. ὑπό and acc., 167, 171, 196; c. acc. 189, 217–18; ἡ ὑποτείνουσα, the hypotenuse, 167, 218
ὑποτιθέναι, to suppose or assume (cf. hypothesis), 196

Χροιά, colour or skin, Pythagorean name for surface, 122
χωρίον, space or area, 206

INDEX OF PROPER NAMES

THE END